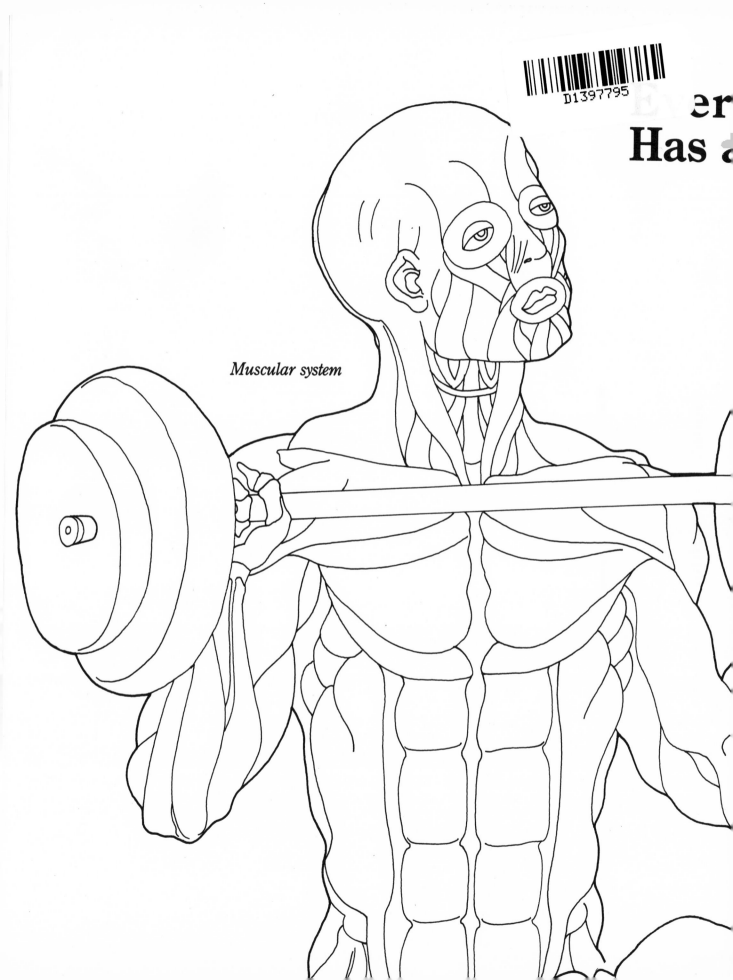

Muscular system

D1397795

body
Body

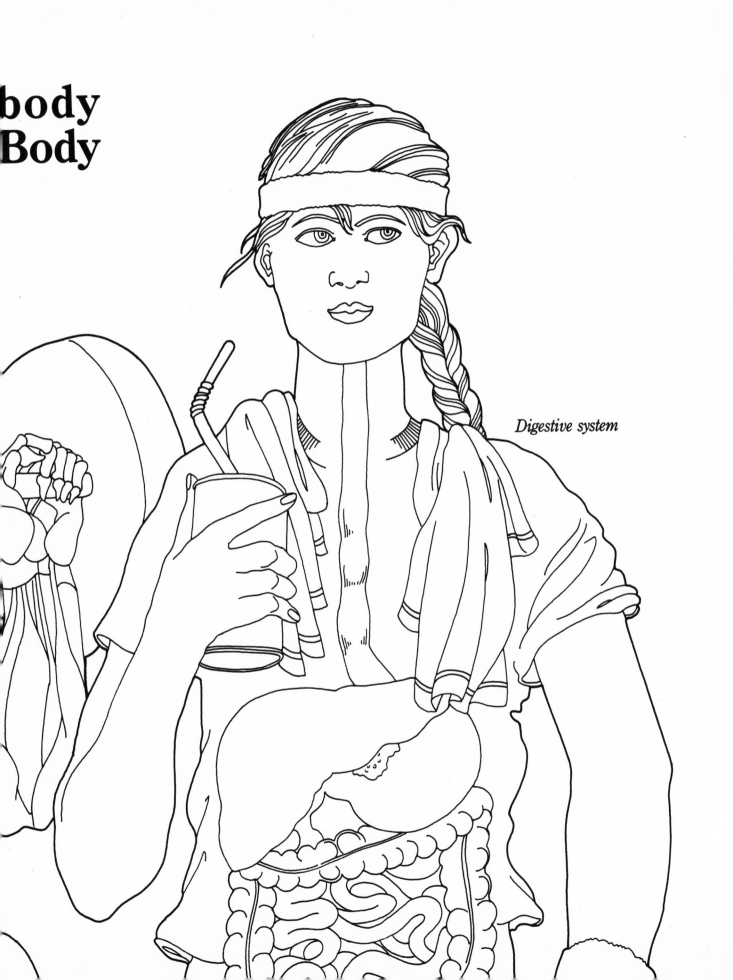

Digestive system

START EXPLORING™

GRAY'S ANATOMY
A FACT-FILLED COLORING BOOK

Freddy Stark, Ph.D.
Original illustrations by Henry Gray
adapted by Helen I. Driggs

RUNNING PRESS
PHILADELPHIA · LONDON

Copyright © 1980, 1991 by Running Press Book Publishers.
Printed in the United States of America. All rights reserved
under the Pan-American and International Copyright Conventions.

This book may not be reproduced in whole or in part in any form or by any means, electronic or mechanical, including photocopying, recording, or by any information storage and retrieval system now known or hereafter invented, without written permission from the publisher.

19

Digit on the right indicates the number of this printing.

ISBN 0-7624-0944-4

Cover design by Alicia Freile
Interior design by E. Michael Epps
Cover, interior, and poster illustrations by Henry Gray,
adapted by Helen I. Driggs
Poster copyright © 1991 by Running Press Book Publishers
Typography by Commcor Communications Corporation,
Philadelphia, Pennsylvania

This book may be ordered by mail from the publisher.
Please add $2.50 for postage and handling.
But try your bookstore first!
Running Press Book Publishers
125 South Twenty-second Street
Philadelphia, Pennsylvania 19103–4399.

CONTENTS

INTRODUCTION

What is anatomy?

Doctors will tell you that anatomy is the study of the bones, muscles, blood vessels, and organs of living creatures. They will also tell you about the excitement and the feeling of discovery that come from seeing how the parts of the body fit and work together.

And what better way to discover the wonders of the human body than by coloring in a selection of drawings from *Gray's Anatomy*, the classic reference book in the field? The *Gray's Anatomy Coloring Book* gives you a chance to learn about the human body and have fun at the same time.

You can choose any colors that please you. But the odds are that if a doctor found some of these colors in your body, he or she would be far from pleased! That's because the color of the organs in the human body provides important clues about the presence of disease. For example, a yellow liver is full of fat, a red liver is full of blood, and a green liver is full of bile. Some livers are multicolored!

Most of your internal organs are shades of tan or pink. Your liver and spleen are purplish-brown, your adrenal glands are bright yellow, and fat is slightly less yellow. The outside of your brain is light gray, and the inside is almost white. Blood, of course, is red. Those veins you thought were blue only seem that color because you're looking at them through your skin.

It's handy to know these colors, but that's no reason why you should feel you have to stick to them. If you want green lungs, go ahead and have them. Explore! Have fun! Discover the wonders of the machine called the human body.

—Jay F. Schamberg, M.D.

GRAY'S ANATOMY

A FACT-FILLED
COLORING BOOK

Heads Up!

You may see skulls and skeletons in scary movies or on Halloween, but there's really nothing spooky about them. Everyone—from the youngest baby to the oldest person—has a skeleton.

Your skeletal system supports your body in the same way that steel girders support a building. It's a protective frame that's both light and strong. If you weigh 85 pounds, your bones weigh only about 15 pounds. Steel bars of the same size might weigh five times as much!

Bones are made mostly of the mineral calcium, the mineral phosphorus, a fiber-like protein called collagen, and water. The minerals give bones strength, and the collagen makes them flexible.

Some bones protect soft tissues and organs. The portion of the skull that protects the brain, the *cranium* (KRAY-nee-um) is composed of thick, interlocking bony plates. These plates are called the *frontal, sphenoid* (SFEE-noyd), *temporal* (tem-POR-ul), *parietal* (pah-RY-eh-tul), and *occipital* (ok-SIP-ih-tul) *bones*.

In adults, these bones meet at joints called *sutures*. In babies, the sutures aren't connected to each other. This gives the brain room to grow. The sutures become rigid and fused as a child grows older.

A side view of the skull showing its major bones. Your skull provides excellent protection for your brain.

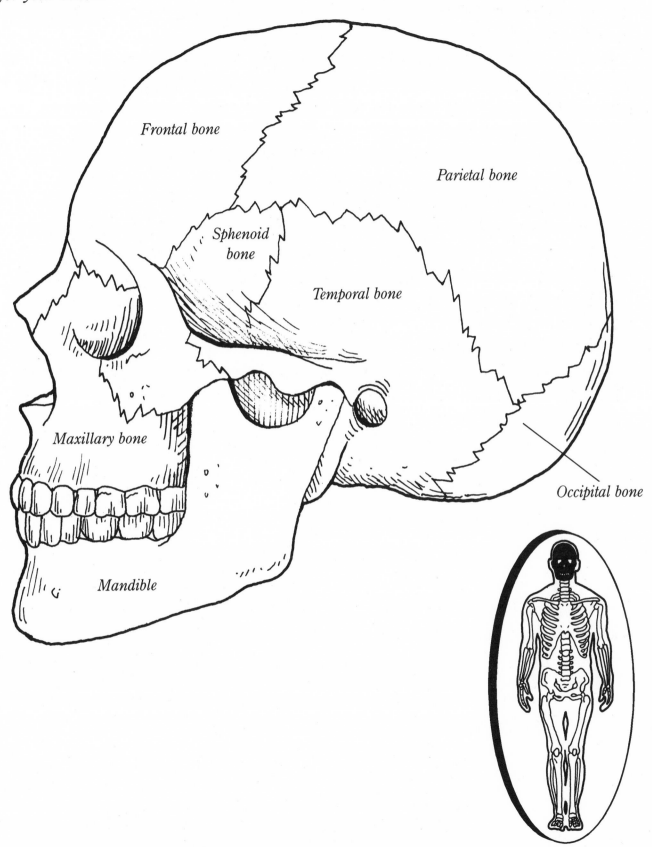

Frontal bone

Parietal bone

Sphenoid bone

Temporal bone

Maxillary bone

Occipital bone

Mandible

Holes in Your Head?

Your jaws are made of two bones. The upper jawbone doesn't move. It's called the *maxillary bone* or *maxilla*. The lower jawbone, or *mandible*, does move. It allows you to eat and speak.

Your upper jawbone has space known as the *maxillary sinus*. The sinus is truly a hole in your head! But don't worry—everybody has one. The sinus acts as a chamber that helps make your voice louder. It's also the main cause of sinus problems.

When you have a cold or flu, your sinus fills with fluid. The sinus's drainage hole isn't on the bottom of the sinus, so the fluid can't totally drain, and you get that "stuffy head" feeling.

Jawbones are unique because of the *teeth* that grow within the bone and poke out, or erupt, into your mouth. The exposed part of the tooth is covered by *enamel*, the hardest substance in your body.

Below the gum line and including the roots, teeth are covered with a softer substance known as *cementum*. Your teeth are attached to your jaws by tough connective tissue.

Most children have 20 teeth, and most adults have 32. Teeth have four specialized jobs. You can tell what their jobs are by studying their shapes.

The teeth at the front of your mouth are shaped like little chisels. They're called *incisors*. They allow you to nibble relatively hard foods such as apples and pears. The next set are the *canines*. Their pointy shape tells you they're used for biting. The *bicuspids* or *premolars* are next. Their primary task, along with the *molars*, is to grind up food with their broad, flat tops.

The last or third molars are also known as *wisdom teeth*. These teeth usually erupt when a person is about 20 years old. Hundreds of thousands of years ago, humans had longer jaws, and so there was space for the wisdom teeth to grow. But today, many people's jawbones have no room for these teeth. If wisdom teeth erupt at a funny angle they must be removed by a dentist.

The left side of the skull, showing baby teeth and permanent teeth. Teeth are shaped for the specialized jobs they do—biting, grinding, or chewing.

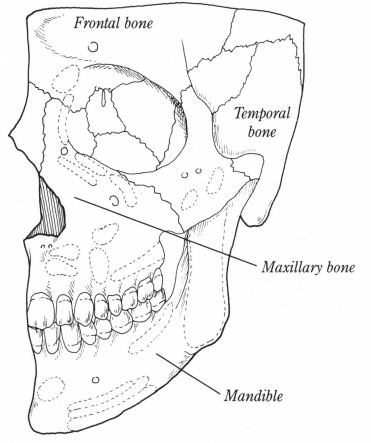

Frontal bone

Temporal bone

Maxillary bone

Mandible

Baby Teeth

Incisors   Canines   Molars

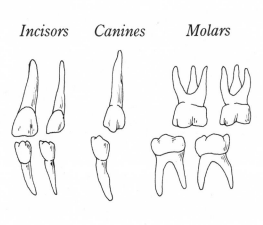

Adult Teeth

Molars   Bicuspids   Canines   Incisors

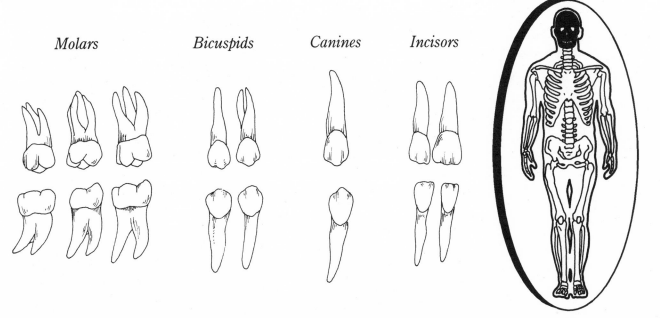

Open Wide

The largest and strongest bone of your face is the *mandible*, or lower jawbone. It forms a hinge joint with the temporal bone of the cranium, allowing you to open and close your mouth.

The mandible also provides a place for the muscles of chewing to attach themselves. The strongest and largest of these muscles is called the *masseter.* Place your fingers on your cheek and clench your teeth. You should be able to feel the masseter because it lies just beneath the skin.

Although the mandible joint is very flexible, it can be knocked out of line by a sharp blow. This is called a dislocation, and when it happens, a person can't close his or her mouth.

Looking at the side of the mandible, you can see a hole called the *mental foramen*. Normally, when we see the word ''mental,'' we think of the brain. But in this case, the word comes from the Latin word *mentum*, meaning chin, because the hole is just behind the point of your chin. Nerves and blood vessels travel through this hole to supply feeling and blood to the chin and lower lip.

As you get older, your mandible bone can wear down, changing its shape. An elderly person's mandible may no longer hold the teeth properly. The mental foramen may shift from its usual position at the chin into the area just below the mouth. This can irritate the nerves which pass through the hole and can cause pain while chewing or wearing dentures.

Your lower jaw changes throughout your life. From top to bottom are the jaws of a baby, a teenager, an adult, and an older person.

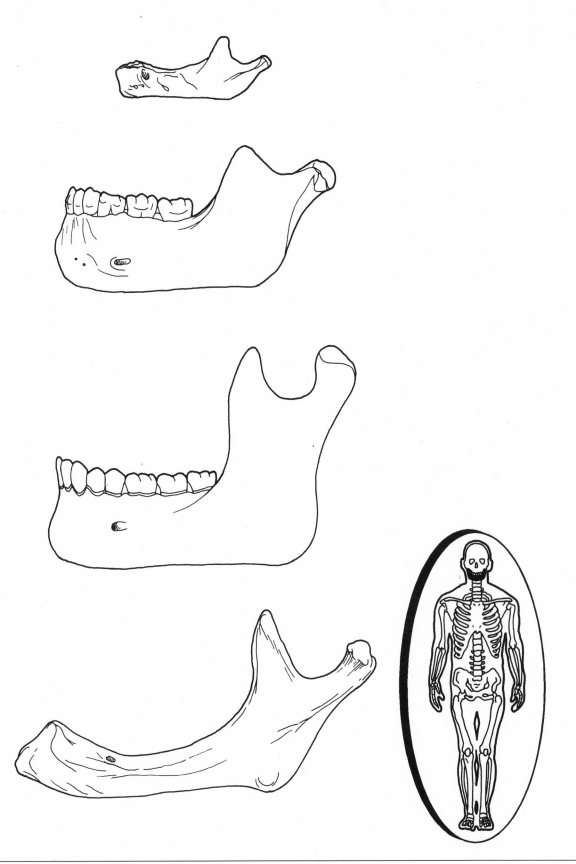

The Roof of Your Mouth

The dome-shaped roof of your mouth and the floor of your nasal cavities are formed by the *palate* (PAL-it). Your palate has two parts: a bony *hard palate* at the front of your mouth, and a fleshy *soft palate* at the back.

The hard palate contains all of your upper teeth, and it's the spot that gets burned when your pizza is too hot. It's formed by a portion of your upper jawbone and two *palatine* (PAL-uh-tyn) *bones*. Like other bones in your skull, the palatine bones have holes in them for nerves, arteries, and veins to pass through.

Your soft palate is behind the hard palate. It's fleshy and doesn't have a bone. The soft palate moves up when you swallow, to prevent food from entering your nasal cavities.

Together with your teeth, tongue, lips, and jaws, your palate is very important for forming words. Your palate also helps your tongue prepare food for swallowing. As you chew, your tongue pushes food against your palate, forming the food into a shape that you can easily swallow.

The top figure shows the hard palate and the upper teeth. The bottom drawing shows the jawbone, or mandible. The masseter muscle attaches to the area on the right of the mandible.

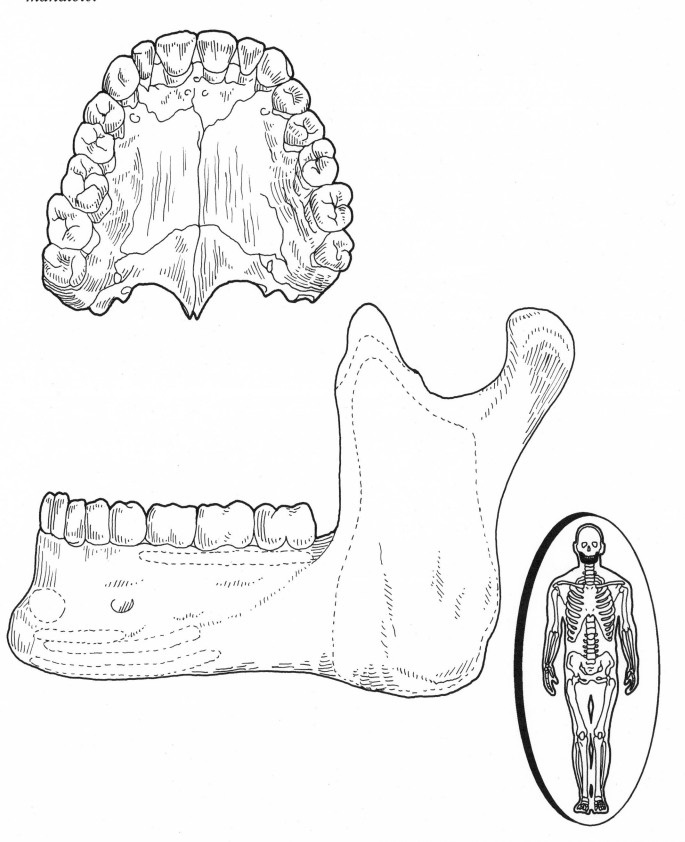

Soft Spots and Strong Supports

The upper part of this drawing shows a newborn baby's skull. The joints between the skull bones, called *sutures*, have not joined and are free to move around. This allows the baby's head to squeeze through the mother's birth canal as the baby is born. The loose joints also allow the baby's brain to grow inside the skull.

Although the adult skull and this drawing of the baby's skull have the same bones, some parts are different. The large, shaded areas in the top and side views of the skull are soft spots called *fontanelles* (FON-tah-nels). These soft spots and the unjoined sutures make a baby's skull more delicate than an adult's. As a baby grows, the sutures join and the fontanelles harden into bone.

The lower part of the drawing shows the base of an adult skull, just behind where the neck joins the skull. This bone is called the *occipital* (ok-SIP-ih-tul) *bone*. The muscles that support and move your head are attached to it. The occipital bone has a large hole in it called the *foramen magnum*. The spinal cord passes through this hole and attaches to the lower part of the brain, called the *brain stem*.

On either side of the foramen magnum, bony knobs allow the skull to rest on the spine. They form a joint with the top neckbone, also called the first *vertebra*. Just above this joint are holes through which the internal jugular veins pass. These veins drain blood from the brain.

The top drawings show two views of a newborn baby's skull. The bones haven't completely joined. The bottom drawing shows the outside of the occipital bone at the base of the skull.

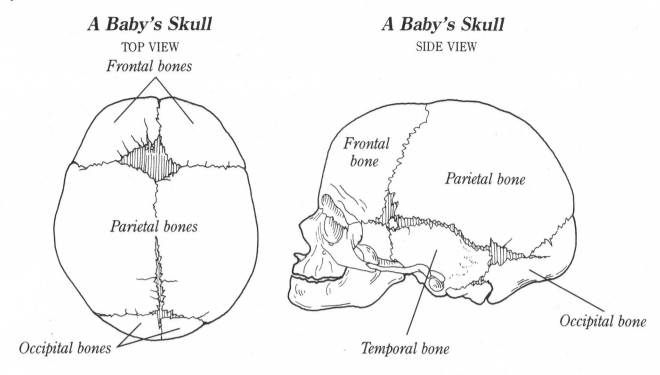

A Baby's Skull
TOP VIEW

Frontal bones

Parietal bones

Occipital bones

A Baby's Skull
SIDE VIEW

Frontal bone

Parietal bone

Occipital bone

Temporal bone

An Adult Skull
BOTTOM VIEW

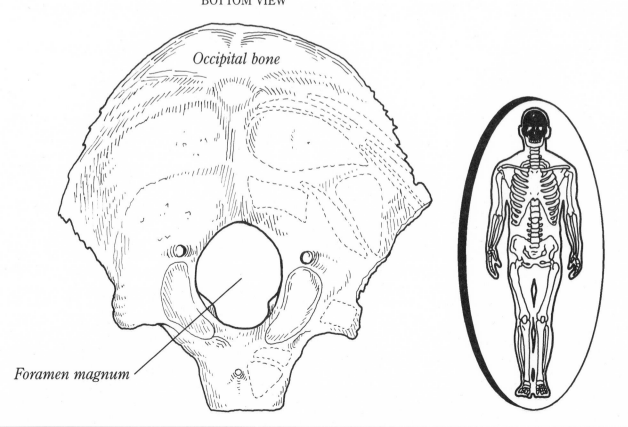

Occipital bone

Foramen magnum

Tough but Fragile

In this view underneath the skull, you can see the *occipital* (ok-SIP-ih-tul) *bone*, the *hard palate*, and the *maxillary teeth*. The dotted lines show many of the muscles responsible for moving your head and jaw.

On the right, the drawing shows the *cheekbone* or *zygomatic arch*. The most powerful chewing muscle, the *masseter*, is attached to this bony arch and the mandible.

The *foramen magnum* is the largest of many holes in your skull. These holes allow nerves, arteries, and veins to pass through the skull. The bones found at the base of the skull can break or fracture because of these holes, and also because these bones are thinner than the rest of the skull. These fractures may affect the brain and are serious, indeed!

At the left of the drawing, you can see the side view of the *upper jaw* or *maxillary bone*, along with the upper set of *teeth* and the *nasal space* above them. Since the maxillary bone is thin, it's among the easiest bones to break. Such fractures can loosen the teeth or make them fall out.

To the left is a side view of the upper jaw, along with teeth and the spaces for the nose and an eye. On the right is the base of the skull as seen from below. The spinal cord fits into the large hole in the center, called the foramen magnum. Nerves and blood vessels pass through the other holes.

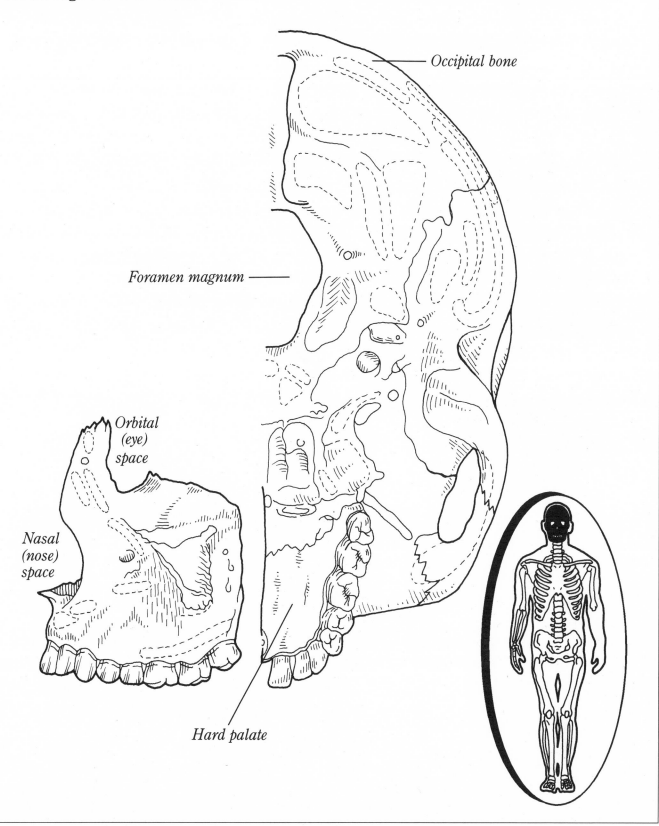

Occipital bone

Foramen magnum

Orbital (eye) space

Nasal (nose) space

Hard palate

Built-in Shock Absorbers

Many of the structures that the brain needs to do its work pass through the base of the skull. Some of the more important structures are the large veins of the brain.

These large veins, called the *venous sinuses* (VEE-nis SY-nus-is), are not in the brain, but in the tough covering of the brain. So the brain has both a hard, bony cranium and a tough tissue covering for protection much in the same way we wear layers of clothing for protection. The *cranial nerves* pass through both these protective coverings.

The skull sits on top of the *spine* or *vertebral column* which has 26 ring-shaped bones. (The spine is often called the *backbone*, even though it's made of many bones.) These bones are called *vertebrae* (VER-tuh-bray), and they protect the *spinal cord*.

The center of the vertebral column has a space for the spinal cord, which has 27 pairs of *spinal nerves*. These nerves carry messages between the spinal cord and the brain.

Between each *vertebra* is a *disc* of shock-absorbing cartilage. Cartilage is a tough tissue found in your joints. These discs are similar to shock absorbers in cars. Just as shock absorbers smooth out the bumps and ruts in the road, these discs absorb the shock of every step you take when you walk or run. Discs can sometimes break and put pressure on the spinal nerves. This causes severe back pain and is called a "slipped disc."

You can see the thick bone that protects your brain in this inside view of the base of the skull. To the right is the spine, made of 26 bony rings.

The Base of the Skull

INSIDE VIEW

Frontal bone

The Spinal Cord

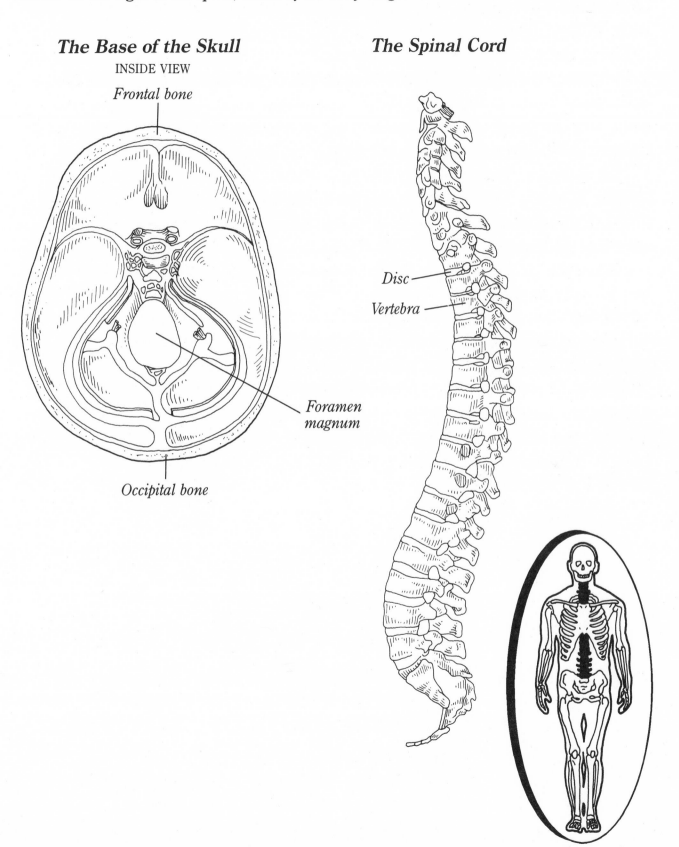

Disc

Vertebra

Foramen magnum

Occipital bone

The Blink of an Eye

When muscle contracts, it moves bones or other body parts. Your body has more than 600 muscles. The smallest ones anchor your tiny ear bones and eardrums, and the largest ones move your legs. Muscles that pull on your skin and lips make you smile and frown.

Muscles usually work in pairs. One will pull forward and the other will pull back. This simple system allows you to make all of your movements—from throwing a ball to threading a needle. Muscle movement is controlled by the brain and spinal cord through their system of nerves.

These drawings show some muscles in and around the skull. The upper drawing shows a large muscle covering the *temporal bone*. It's called the *temporalis muscle*. This muscle is attached to the mandible and, together with the masseter muscle, is one of the muscles you use for chewing. If you place your fingers on your temple and clench your teeth, you can feel this muscle contracting.

The bottom drawing shows the *socket* or *orbit* of the eye. Each eye has six small muscles that move the eyeball. You can see one of these muscles just above the eye, traveling through a tiny pulley called the *trochlea* (TROK-lee-uh). This pulley allows the muscle to move the eye diagonally.

Another muscle in the orbit opens the *upper eyelid* when it contracts. A different muscle closes the eyelid. It takes only about one-tenth of a second to blink your eyes. While you're awake, you blink about 10,000 times. Each blink keeps your eyes moist and wipes away dust.

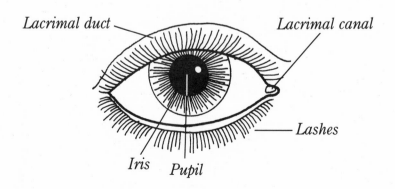

Tears are produced in the lacrimal duct.
They soothe the eye, then they drain
through the lacrimal canal.

One muscle that helps you chew is the temporalis muscle. It's located on both sides of your skull. Pictured at bottom is an eyeball in its socket, along with the muscles that move the eye.

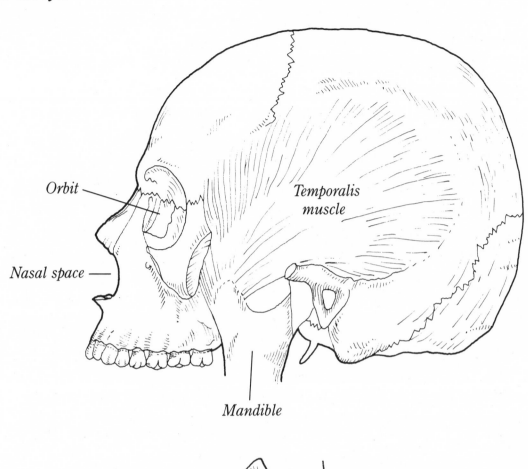

Orbit

Nasal space

Temporalis muscle

Mandible

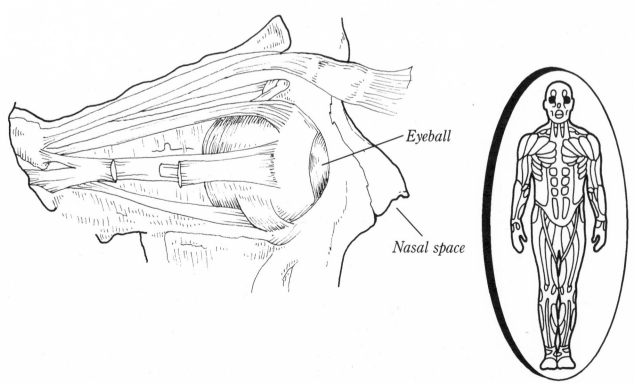

Eyeball

Nasal space

Put on a Happy Face

Think about all the faces you make in one day. You smile with pleasure, curl your lip in disgust, raise your eyebrows in surprise and on and on. These little gestures require the surface muscles of your head, face, and neck.

You have two kinds of muscles in your face—long, straight ones covering your neck and cheeks, and circular ones around your eyes and mouth. These muscles pull on skin rather than bone.

The muscle covering your neck is called the *platysma* (PLAT-es-ma). It tenses the skin of the neck and makes it easier for men to shave their beards. Other straplike muscles attach to the corners of your mouth. The ones going upward help you smile, and the ones going downward make you frown.

Your cheek muscle is called the *buccinator* (BUK-sin-ay-tor). It presses your cheeks against your teeth, allowing you to sip through a straw. Some trumpet players stretch their buccinator muscles so much that their cheeks balloon out when they play!

Other muscles in the surface layer can move the nostrils or wiggle the ears, but these muscles aren't well developed in everyone.

When the circular muscles around your eyes contract, your eyes close firmly to protect your eyeballs. These muscles also make wrinkles around your eyes when you squint. The circular muscle around your mouth closes your mouth and purses your lips for whistling or kissing. This muscle also helps you form words and sounds. Working together with the buccinator, it helps keep food between your teeth while chewing.

Your face is almost completely wrapped in muscles. These muscles let you change your expression.

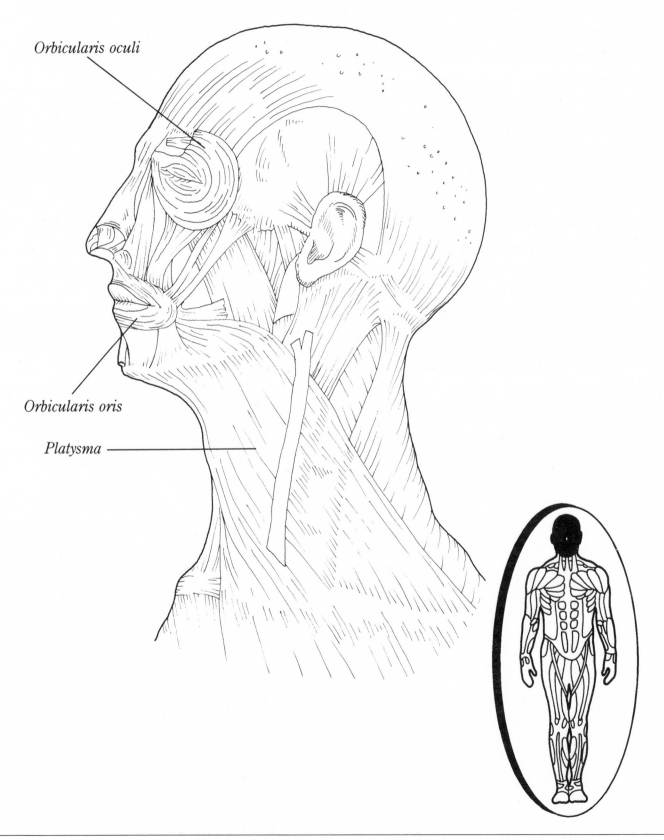

Orbicularis oculi

Orbicularis oris

Platysma

Shrugging Your Shoulders

Beneath the surface muscles is a layer of *deep muscles*. This drawing shows some of the deeper muscles of the neck. The *sternocleidomastoid* (STER-no-KLY-do-MASS-toyd) is a very long word for two very large muscles on either side of your neck. These are the muscles that contract to lift your head off your pillow. When only one side contracts, your head tilts to the right or left.

The muscle at the back of your neck is called the *trapezius* (trah-PEE-zee-us). The trapezius moves your shoulder blades upward when you shrug your shoulders. You can feel the trapezius contract by placing your hand at the back of your neck and shrugging.

The trapezius is a common place for tension and pain. Often when you wake up with a "pain in the neck," the trapezius muscle is to blame.

In the drawing, you can see a triangle between the trapezius and sternocleidomastoid. Several deeper muscles in this area help keep your head upright.

Between the sternocleidomastoid and the front of the neck is another triangle. This area contains the muscles you need for swallowing. These muscles attach either to the underside of the mandible or to the small *hyoid bone* in the upper part of your neck.

Your many neck muscles let you swallow and move your head.

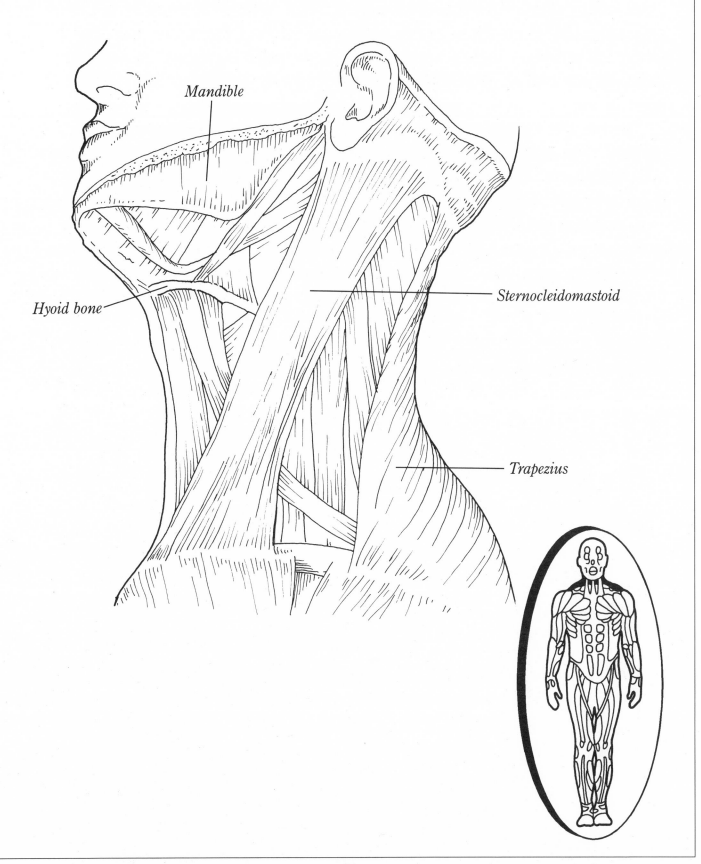

Mandible

Hyoid bone

Sternocleidomastoid

Trapezius

Disease Fighters

Although you may think bone is solid, it isn't. Many bones have holes. The upper drawing shows that bones contain blood vessels. In this case, we can see the veins of the skull, which lie just beneath its hard covering. The reason that bones have blood vessels is that bone is a living tissue and needs a good blood supply to nourish it.

In order to work effectively, blood has a powerful friend in *lymph* (limf). Lymph is a watery fluid that acts as a servant to the cells of the body and blood. It carries oxygen and nutrients from blood vessels to the cells and carries wastes from the cells back to the blood. Lymph also carries the fluid and proteins that can escape from blood back to the circulatory system.

The circulatory system includes the heart, veins, and arteries that carry blood through the body. Tubes that carry lymph are called *lymph vessels*. *Lymph nodes* are large tissue masses in the lymph vessels that help remove dead cells and foreign material.

Disease-fighting white blood cells also gather in lymph nodes. These cells are called *lymphocytes* (limf-O-syts), and they tend to gather in groups to fight infection.

When you bang your arm or leg and it swells up, the swelling is caused by the lymph rushing to the injured area to do its work.

L Y M P H
TO THE
RESCUE

When you cut yourself, germs can get into your blood. But don't worry—your lymph nodes protect you.

Your lymphocytes and other white blood cells travel throughout your body to hunt down and destroy germs. One kind of lymphocyte releases poisons which are harmless to you but deadly to germs. Other kinds of lymphocytes gobble up germs like gumdrops, or hunt down and kill cancer cells.

During an infection, such as the mumps, the lymph nodes can fill with lympho-cytes and swell. This is mistakenly called "swollen glands."

The bones of your skull have veins inside them, as shown in the top drawing.
The drawing at the bottom shows the lymph vessels that protect and nourish the cells
of your body.

Blood Vessels

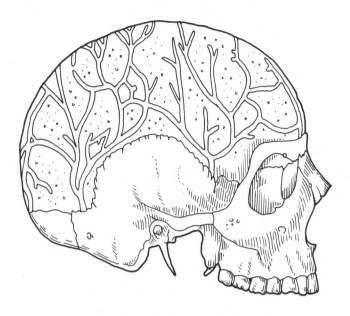

Lymph Vessels

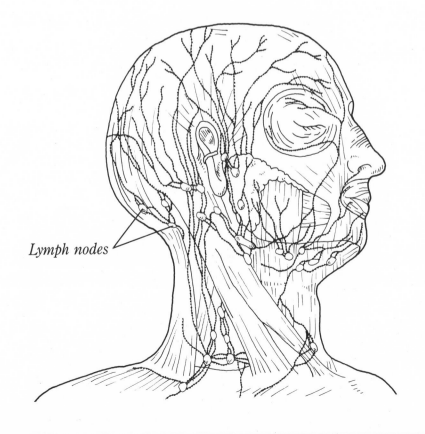

Lymph nodes

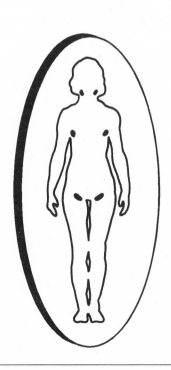

Finding Your Pulse

All the living cells of your body need food and oxygen to stay alive. The *arteries* are the pulsing tubes that carry nutrient- and oxygen-rich blood away from your heart to all parts of your body.

Arteries are lined with muscles that contract. When this happens, the arteries close, routing the blood to where it's needed most. When you run, the arteries in your stomach and intestines contract while the arteries in your legs relax. That lets more blood flow to your hard-working leg muscles.

Unlike the muscles you use to move your arms and legs, you can't control the muscles in your arteries. These muscles are called *involuntary muscles*.

The top drawing shows the main artery that supplies blood to your neck, face, and scalp. It's called the *external carotid* (ka-RAH-tid) *artery.* This artery has many branches. If you press your fingers to your neck just beneath your jaw, you can feel blood pulsing through this artery. Sometimes you can see one of the external carotid's branches pulsing at a person's temples.

The lower drawing shows the major artery of the eye. This artery is a branch of the *internal carotid artery*. It's called the *ophthalmic* (ahf-THAL-mik) *artery*. To the right of the eyeball is the *tear gland* or *lacrimal* (LAK-rih-mul) *gland*. This gland produces the tears to keep your eyes clean and moist.

You can also see the *optic nerve* at the rear of the eye. This nerve carries to the brain the signals that let you see.

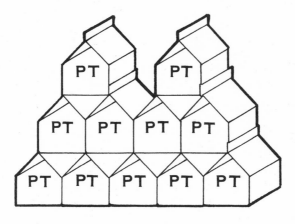

A 150-pound person has about 11 pints of blood flowing through his or her body.

The top drawing shows the muscles of facial expression and some of the arteries that supply blood to your face. The bottom drawing shows the major blood vessels of the eye, seen as if all the surrounding bones and tissues had been removed.

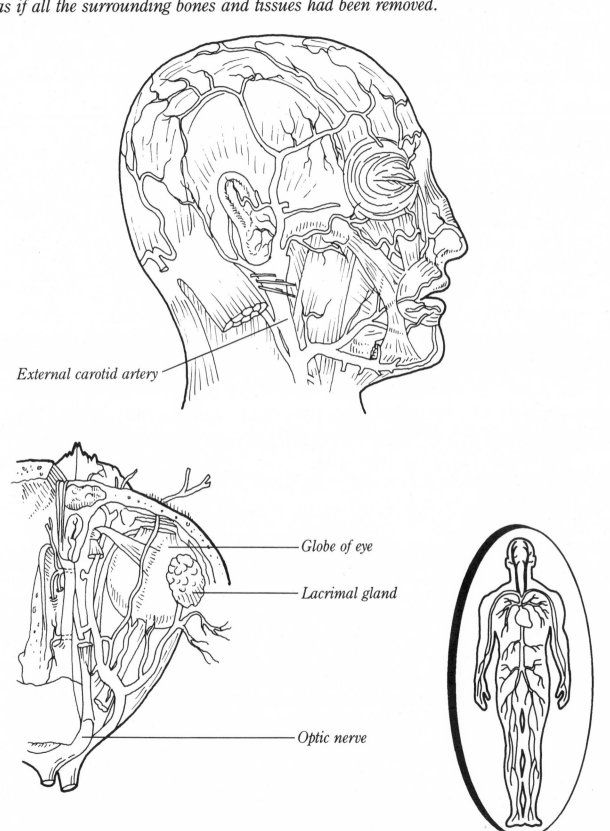

External carotid artery

Globe of eye

Lacrimal gland

Optic nerve

Just Below the Surface

This drawing shows the large branches of the *external carotid artery*. You can see two of the major branches in the head. The *superficial temporal artery* supplies blood to the scalp. This is the artery that pulses at your temples.

The other branch is the *maxillary artery*. This artery travels deep within the face and *maxillary bone*. It supplies blood to the brain's tough covering and to your teeth. Blood from the maxillary artery also supplies the nasal cavities, and through holes in the maxillary bone and mandible, it delivers blood to the facial muscles and skin.

Just behind the maxillary artery are two muscles called the *pterygoid* (TER-ih-goyd) *muscles*. Part of these muscles have been removed from this drawing to give you a better view of the maxillary artery. Together with the *masseter* and *temporalis muscles*, the pterygoid muscles help you chew your food.

Just behind the superficial temporal artery are two other structures. One is a hole in the *temporal bone*. This hole is the opening in the skull for the *ear canal*. The ear canal brings the sounds we hear to the eardrum. The other structure is a thin and pointy bone called the *styloid process*. Some of the muscles that help you swallow and move your tongue are attached to it.

A view of the face showing some of the deeper muscles and arteries.

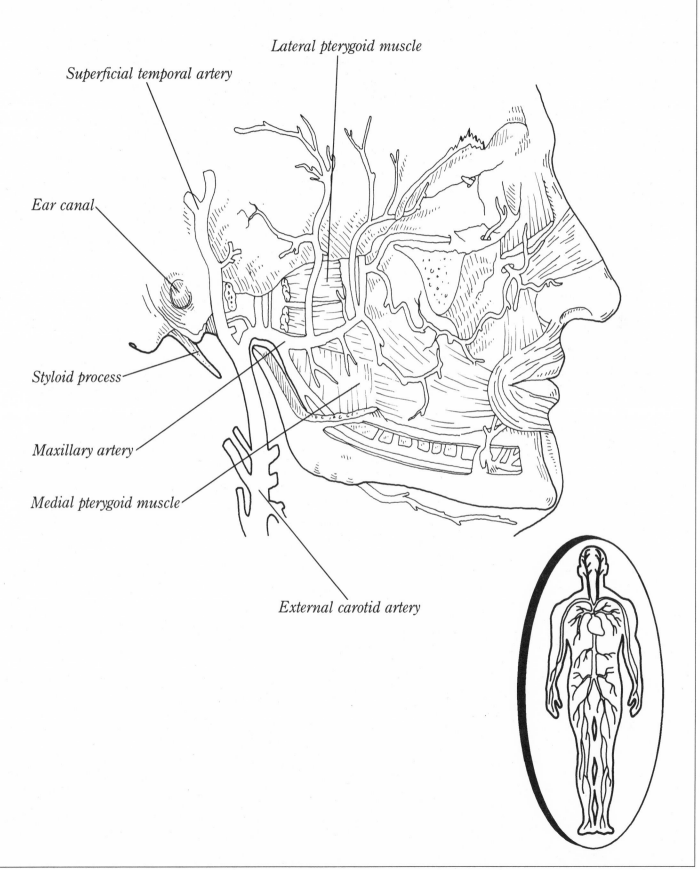

Lateral pterygoid muscle

Superficial temporal artery

Ear canal

Styloid process

Maxillary artery

Medial pterygoid muscle

External carotid artery

Get It into Your Head

Your head receives blood from two major arteries on each side of your neck, called the right and left *common carotid arteries*.

Each common carotid artery branches into two other arteries, the *internal carotid artery* and the *external carotid artery*. The external carotid arteries carry blood mainly to the face, scalp, and neck. The internal carotid arteries carry blood to the brain. To reach the brain, the internal carotid arteries pass through an opening at the base of the skull.

At the place where the carotid arteries branch is a small but vital structure that helps control your blood pressure. It's called the *carotid body*. The control of blood pressure is important because high blood pressure can damage blood vessels. If the blood pressure is too low, the brain and other structures may not get a good supply of nutrients and oxygen.

This drawing also shows a branch of the external carotid artery called the *facial artery*. This snake-like artery travels over the lower jawbone, or *mandible*. You may be able to feel it pulsing at your mandible if you don't press too hard.

The facial artery has many branches to the structures of the face and the back of the mouth.

The right side of the neck, seen with the head turned to the left. Some muscles have been omitted from this drawing so that you can see the main arteries that supply the head and neck with blood.

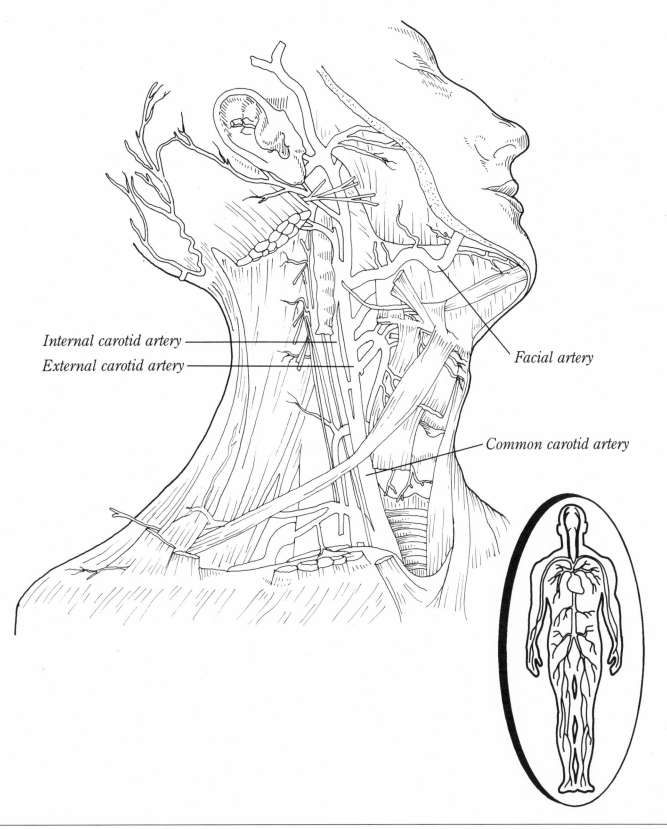

Internal carotid artery

External carotid artery

Facial artery

Common carotid artery

Going for the Jugular

Tubes that bring blood from the body back to the heart are called *veins*. Blood in your veins has less oxygen than blood in your arteries. It also contains wastes from the tissues it passes through. *Surface veins* travel just beneath the skin, and *deep veins* usually travel alongside the deeper arteries.

The veins and arteries in your head and body are like a series of small streets and superhighways. All your blood vessels connect with one another, creating a pathway thousands of miles long if stretched end to end.

This drawing shows the surface veins on one side of the head and neck. The most important surface vein of the neck is the *external jugular vein*. While looking in a mirror, hold your breath and push in your chest and abdomen. You'll be able to see your external jugular vein. This vein has many connecting branches that collect blood from your scalp and face.

The other important neck vein is called the *internal jugular vein*. This deep vein travels alongside the *carotid artery*. It collects blood from the brain, the skin and surface muscles of the face, and the neck. Some of its branches connect with the external jugular vein.

Here are the surface veins of your head and neck, along with the first layer of muscles.

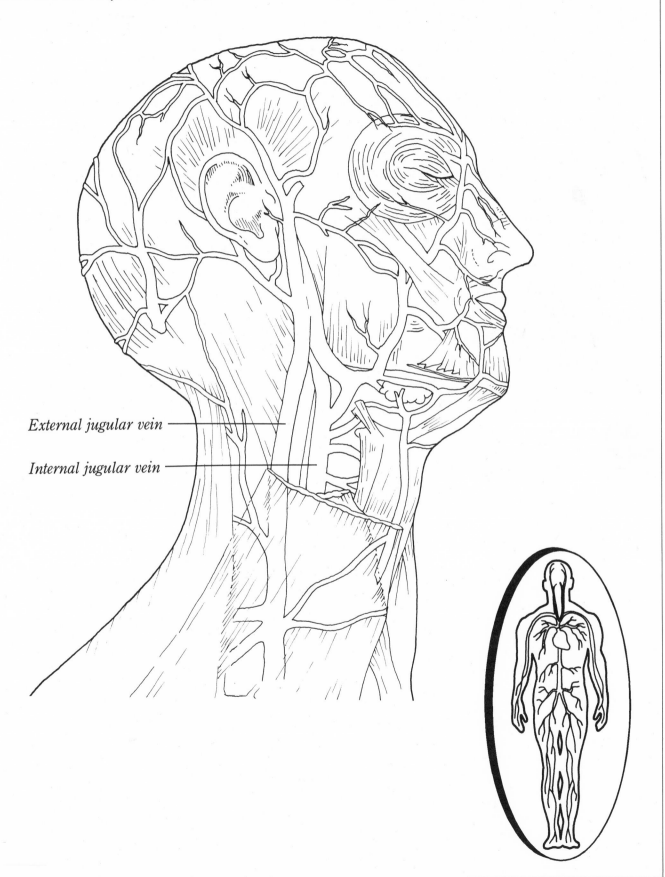

External jugular vein

Internal jugular vein

Easy to Swallow

Your brain requires a steady supply of blood to work properly, and the *internal carotid arteries* and the *vertebral arteries* make sure that the brain gets all the blood it needs. The vertebral arteries also supply blood to the spinal cord in the neck. Once in the skull, the arteries are connected to each other by a circle of smaller arteries.

Look at the throat of the person in this drawing. You can see three structures that help us speak and swallow. The upper structure is the *hyoid bone*. This bone is the upper part of the *voice box* or *larynx* (LAR-inks). Your voice box contains your vocal cords. You might have already guessed that your voice box is where your voice comes from.

The structure below the hyoid bone is the largest piece of cartilage of the voice box. It's called the *thyroid cartilage*, but most people know it as the Adam's apple. You can feel your *Adam's apple* in the center of your neck under your chin.

Below the thyroid cartilage is the *trachea* (TRAY-kee-uh), or *windpipe*. This tube is made of rings of cartilage. It allows air to travel in and out of your lungs. When you swallow, muscles attached to the hyoid bone and thyroid cartilage pull your trachea up and down. You can feel it happening by touching your Adam's apple while swallowing. This motion keeps you from choking by preventing food from entering the larynx and trachea.

Your brain receives much of its blood from the internal carotid and the vertebral arteries. Also shown is the thyroid cartilage, or ''Adam's apple.''

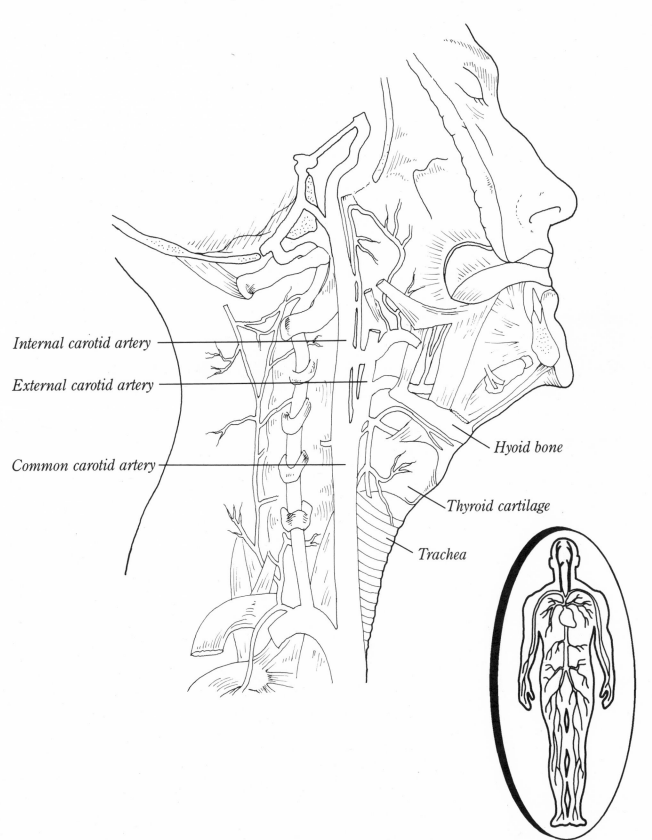

Internal carotid artery

External carotid artery

Common carotid artery

Hyoid bone

Thyroid cartilage

Trachea

A Bundle of Nerves

The strong bones of your skull protect a vital organ—your *brain*. Your brain weighs only about three pounds, but it has many complicated tasks to perform.

Your brain coordinates your movements, controls your breathing, and lets you feel hunger, pain, sadness, and happiness. All the information about the world around you and all the things you do are controlled by your brain through a system of *nerves*.

Sensory nerves carry information to your brain from your five senses: sight, hearing, taste, smell, and touch. When the information from your sensory nerves reaches your brain, your brain sends messages of its own to the body.

For example, if you smell dinner cooking, your brain may make you feel hungry. Or if you see a scary movie, your brain may make your heart beat faster. Nerves that bring messages from your brain to your organs and muscles are called *motor nerves*.

All of the signals going to and from your brain travel through your spinal cord. Together, the brain and spinal cord are called the *central nervous system*.

The Parts of a Nerve Cell

Cell body

Dendrites (gather information)

Axon (carries message)

Everything you do is controlled by this 3-pound wonder—your brain. This is a view of the bottom of the brain.

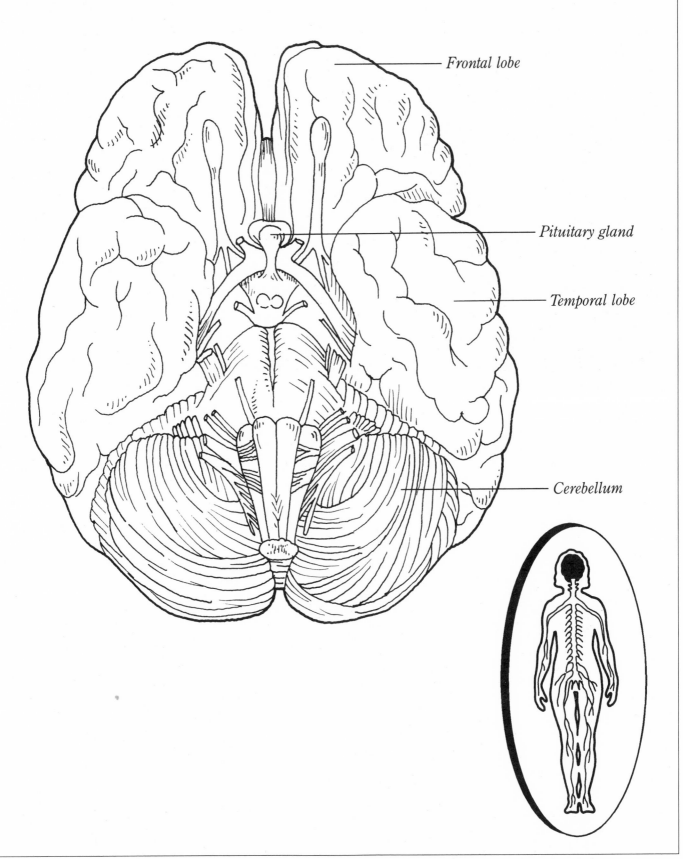

Frontal lobe

Pituitary gland

Temporal lobe

Cerebellum

The Three Parts of Your Brain

The brain can be divided into three parts: the *cerebrum*, the *brain stem*, and the *cerebellum*. The cerebrum is the largest part of your brain. It's made mostly of nerve cells, or neurons. Neurons transmit electrical signals within the brain. These electrical signals can travel as fast as 500 feet per second. When you're thinking or walking, you're using your cerebrum.

Your *brain stem*, which connects your brain to your spine, controls your heartbeat, breathing, and other bodily functions. These functions are called involuntary because they happen automatically—you don't have to remember to do them.

Your brain stem also controls your hormones. *Hormones* are the messengers produced in your glands. Hormones control how tall you grow, how your body uses food, and whether or not you'll grow a beard or lose your hair. The *pituitary* (pih-TOO-ih-ter-ree) *gland* is the part of the brain stem that controls the other hormone-making glands in your body. It's sometimes called "the master gland."

Your *cerebellum* (ser-uh-BELL-um) is located at the back of your brain behind the brain stem. One of its main tasks is to coordinate the contraction of your muscles so you can keep your balance.

A BIG YAWN

When you breathe, your lungs take in oxygen and give out carbon dioxide. When you're tired, your body slows down and doesn't get rid of all the carbon dioxide. Your brain sends you the signal to yawn, and before you know it, you're taking in a huge breath of oxygen-rich air and clearing out the carbon dioxide.

Sometimes, seeing someone else yawn (or just thinking about yawning) can make you yawn.

You can see the major parts, or lobes, of the brain in this view of its right half.

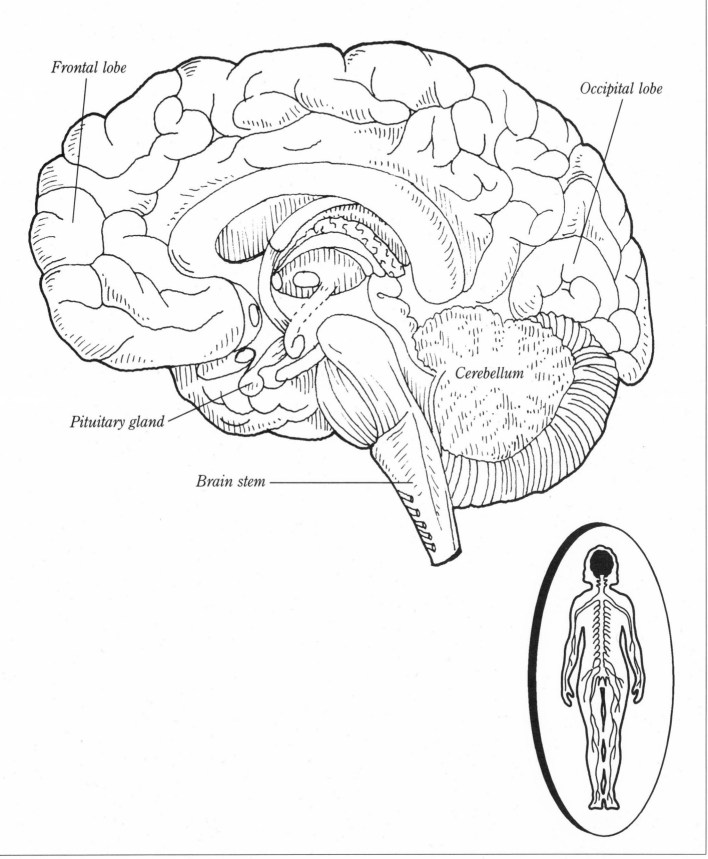

Frontal lobe

Occipital lobe

Cerebellum

Pituitary gland

Brain stem

Thinking about Thinking

The brains of fish, frogs, chickens, and humans have similar parts. How intelligent an animal is depends on the size and complexity of the *cerebrum* (suh-REE-brum).

The drawings show a human cerebrum's four major parts, or lobes: the *frontal*, *parietal* (pah-RY-eh-tul), *occipital* (ok-SIP-ih-tul) and *temporal* (tem-POR-ul) *lobes*. Each lobe has its own duties to perform.

The *frontal lobes* are located at the front of the brain. It's here that all the complicated movements involved with speaking are controlled. A section in the frontal lobe called *Broca's speech area* coordinates the palate, lips, tongue, and other structures that enable you to speak.

Each part of your body is connected to two groups of nerve cells, or neurons, in the frontal lobe. The first area receives the signals of sensation such as hot, cold, and touch from the one side of your body. The neighboring area sends back signals to the muscles on that side of your body.

The tasks of the *parietal lobes* are very complicated. They help the brain to understand and react to all the sensory signals coming to your brain from your body.

The *temporal lobe* helps you understand speech. It's connected by neurons to Broca's area in the frontal lobe. The temporal lobe also allows you to hear and to keep your balance.

The *occipital lobe* is at the rear of the brain. It's often called the visual cortex because it's the part of the brain that enables you to see.

The bottom drawing shows a striped area between the two cerebral hemispheres. This area connects the two hemispheres and is called the *corpus callosum*. This connection allows the left and right sides of the brain to "talk" to each other and to coordinate their tasks.

The largest part of your brain is the cerebrum. The top drawing shows the major arteries supplying blood to the cerebrum. The bottom drawing shows its outer surface as seen from the top.

SIDE VIEW

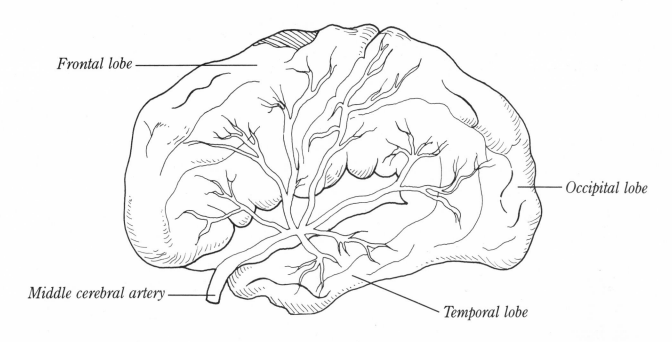

Frontal lobe

Occipital lobe

Middle cerebral artery

Temporal lobe

TOP VIEW

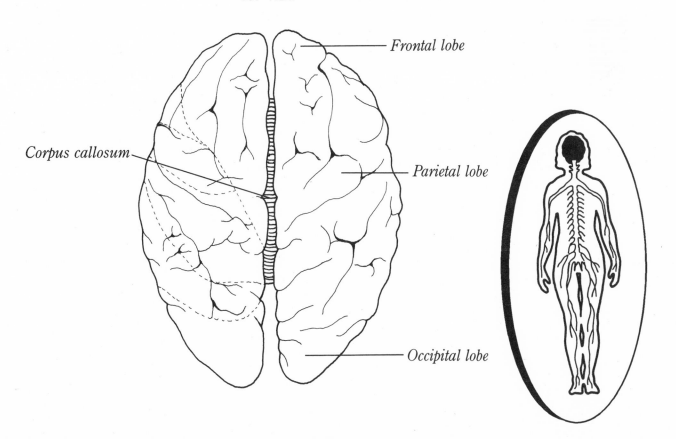

Frontal lobe

Corpus callosum

Parietal lobe

Occipital lobe

Not Just a Pretty Face

To feel a snowflake on your nose, to watch a tennis game, or to smell a rose, you use 12 pairs of nerves in your brain.

Your sense of smell comes from the *first cranial nerve*; your sense of vision comes from the *second cranial nerve*; and the *third*, *fourth*, and *sixth cranial nerves* control the small muscles of your eyeballs.

This drawing shows the *fifth cranial nerve* or *trigeminal* (try-JEM-ih-nul) *nerve* and its branches. You can see some of its branching nerves pass from the skin through holes in the facial bones, and back to the trigeminal nerve.

The trigeminal nerve is the main sensory nerve of the face. When you feel your fingers touching your cheek or cold water splashing on your face, it's the trigeminal nerve that carries these sensations to your brain. The trigeminal nerve also supplies sensation for your tongue, teeth, the inside of your mouth, and your nasal cavities.

Your facial muscles are controlled by the *seventh cranial nerve*, called the *facial nerve*. A branch of the facial nerve, the *chorda tympani*, supplies your sense of taste. The chorda tympani travels past the *eardrum*, which is also called the *tympanic membrane*.

In this drawing, you can see the chorda tympani joining with the branch of the trigeminal nerve leading toward the tongue.

In this see-through view of the side of the head, you can see the branches of the trigeminal nerve. It supplies most of the feeling in your tongue, teeth, nose, and mouth.

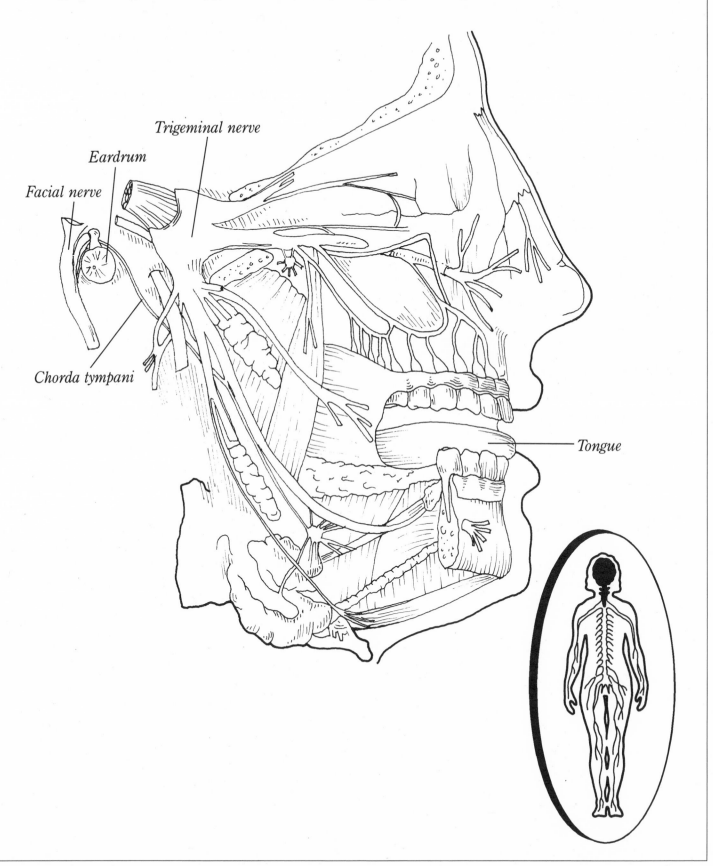

Trigeminal nerve

Eardrum

Facial nerve

Chorda tympani

Tongue

Do You See What I See?

Your *eyes* are your windows on the world. These amazing organs not only allow you to see in color, but they also allow you to see in the dimmest light.

Light enters the clear *cornea* at the front of the eye. After passing through the watery *aqueous humor* (AY-kwee-us HEU-mur), light passes through the *pupil*. The pupil is the black part of the eye. It's really nothing more than a hole in the *iris*, the colored part of the eye. The iris is a circular, involuntary muscle that controls the amount of light entering your eye. In bright light, the iris contracts, making the pupil very small. You can see this happen if you look into a mirror and shine a light in your eyes.

Light is focused by the *lens*. The lens can change its shape, depending on the distance of the object you're looking at. The lens focuses the light on the *retina* (RET-ih-nuh). The retina contains cells called *photoreceptors* (FO-tow-REE-sep-tors) that change light into electrical signals. These signals travel along the neurons of the *optic nerve* (the second cranial nerve) to the brain.

The gel-like *vitreous humor* (VIT-ree-us HEU-mur) helps the eye keep its round shape. If the shape of the eyeball changes just a bit, the lens won't be able to focus light on the retina correctly. When this happens, a person's vision gets blurry and requires glasses or contact lenses to correct it.

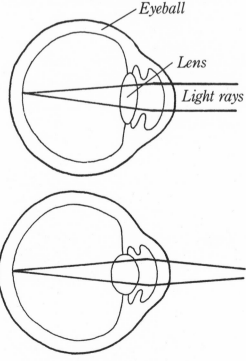

Your lenses change shape to focus on close or distant objects.

Three views of the eye, showing its major parts. The bottom drawing shows a view of the inside.

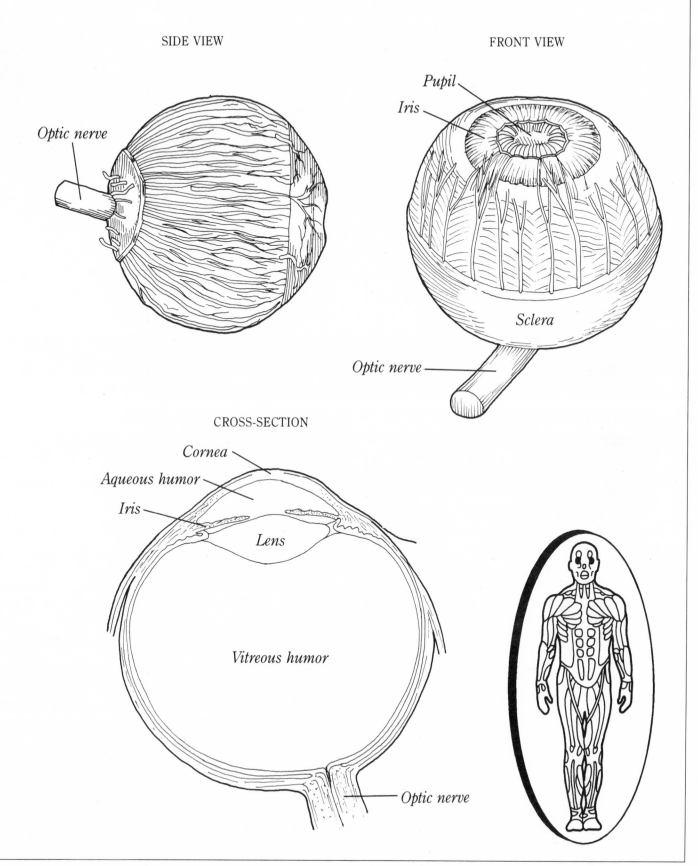

SIDE VIEW

Optic nerve

FRONT VIEW

Pupil

Iris

Sclera

Optic nerve

CROSS-SECTION

Cornea

Aqueous humor

Iris

Lens

Vitreous humor

Optic nerve

You Don't Have to Shout!

Whether it's a pin dropping or the roar of a jet plane, your ears are sensitive to a variety of sounds.

Sound is caused by vibrations that create waves of pressure in the air. Your ear changes these pressure waves into electrical signals that travel along the *auditory nerve* (the eighth cranial nerve) to your brain.

The outermost part of the ear is called the *pinna*. It helps to funnel sound down the *ear canal* to the *eardrum*. The eardrum, or *tympanic membrane*, carries the vibrations to the three smallest bones of your body. These bones are known as the hammer, anvil, and stirrup because that's what they resemble. Together they are called the *ossicles*, and they're in your middle ear.

The ossicles raise the energy of the vibrations more than ten times. Then they pass the vibrations to the snail-shaped *cochlea* (KOK-lee-uh). The cochlea changes the vibrations into electrical signals and sends them to the brain.

The pinna is made of a flexible cartilage. However, the delicate eardrum, ossicles, and cochlea are protected by the hard *temporal bone* of the skull.

The bottom drawing shows the eardrum without the ossicles. You can see the *facial nerve* (seventh cranial nerve) and the *chorda tympani*. Another structure in this drawing is the *Eustachian* (yoo-STA-shun) *tube*. This tube connects your middle ear with the uppermost part of your throat. It helps keep the pressure equal between both sides of your eardrum. That helps you keep your balance. You can feel it work when you ride a roller coaster or an airplane and your ears pop.

It's easy to recognize the outer ear, also called the pinna. Inside the ear are the smallest bones in your body, called ossicles. You can see them in the middle picture, which shows a side view of the ear. The bottom picture shows the eardrum and Eustachian tube.

Outer Ear

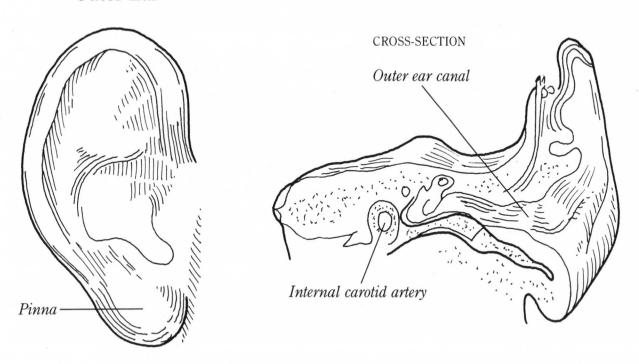

CROSS-SECTION

Outer ear canal

Internal carotid artery

Pinna

The Middle Ear

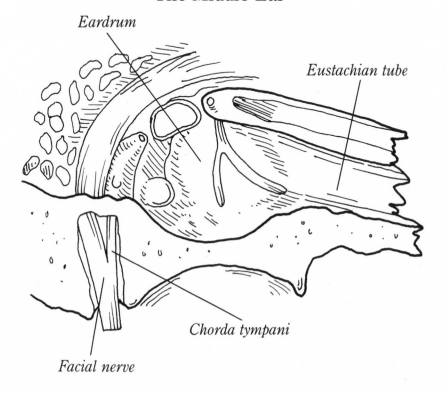

Eardrum

Eustachian tube

Chorda tympani

Facial nerve

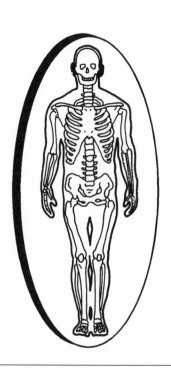

What's That Smell?

Have you ever entered a freshly painted room, a bakery, or some other place filled with a strong aroma? After staying in that room for just a few minutes, your nose gets used to the smell and you no longer notice it.

Your nose is always sniffing out new smells in order to tell you more about what's going on around you. By adjusting to a lingering fragrance, it makes itself more sensitive to newer odors that may be in the air.

Your sense of smell depends on your *nose* and the *olfactory nerves* (the first cranial nerve).

The olfactory nerve is in the upper part of your nasal cavity, or nose, where cells detect odors. The cells change the odors into electrical signals, and the nerve carries these signals to the brain.

Your sense of smell is closely tied to your sense of taste. Try holding your nose shut while eating. You may notice that your food loses much of its taste. Taste and smell are known as our *chemical senses* because they can detect certain chemicals in the air and in our food.

Your nose is made of flexible cartilage and harder bone. The *nasal septum* is cartilage, and it may move from its normal position in the center to one side. This can cause breathing difficulties and snoring. Behind the nasal septum are two bones, the *ethmoid* and the *vomer*. Both bones are thin and can fracture if hit hard.

The outer part of the nose is made of the two *nasal bones* and cartilage around the nostrils. The inside of the nose has a moist membrane and good blood supply. This helps warm and moisten the air you breathe.

Your nose is a combination of cartilage and bone. The top left drawing shows the outside. On the top right are the delicate bones separating the two sides of the inside of your nose. At the bottom is a view of nostrils as seen from below.

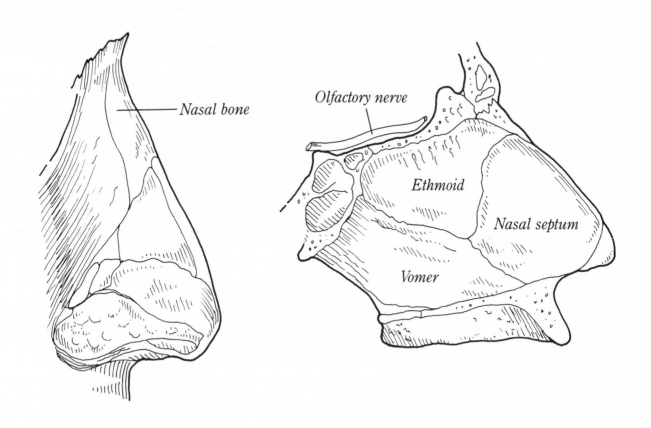

Nasal bone

Olfactory nerve

Ethmoid

Nasal septum

Vomer

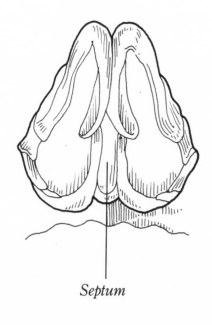

Septum

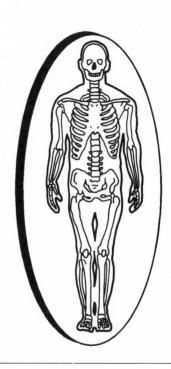

Taste Testing

Everyone knows that you use your tongue to taste food, but did you know that different parts of your tongue are sensitive to different tastes?

You taste sweet and salty foods mainly at the tip of your tongue, sour foods at the sides, and bitter foods at the back.

The nerve that supplies your sense of taste is the *chorda tympani*, a branch of the *facial nerve* (the seventh cranial nerve). It's not the only nerve in your tongue. Your tongue is also very sensitive to touch, heat, and cold. These feelings come mostly from the *trigeminal nerve* (the fifth cranial nerve).

The left drawing also shows two other structures. The *uvula* is the part of the soft palate that hangs down at the back of your mouth and wiggles when you breathe. It helps form sounds in some languages, but most of us have no use for the uvula.

Your *tonsils* are a part of your lymph system, which acts as a filter to trap bacteria or germs that can make you ill. Sometimes your tonsils can get badly infected while trying to protect you. Then they must be removed.

A TASTEFUL EXPERIMENT

Try this experiment with a friend: Get three saucers and some cotton swabs. Put sugar water in one saucer, salt water in another, and diluted lemon juice in the third.

Ask your friend to cover his or her eyes and nose. Dip a swab into one of the solutions and touch it to one spot on your friend's tongue. Ask your friend to guess the taste.

Repeat this experiment with all three flavors on different parts of your friend's tongue. Your friend should sip some water between trials to clear his or her tongue.

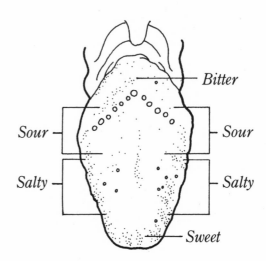

Different areas of your tongue are sensitive to different tastes.

This picture of the tongue also shows a tonsil at the back of the throat.

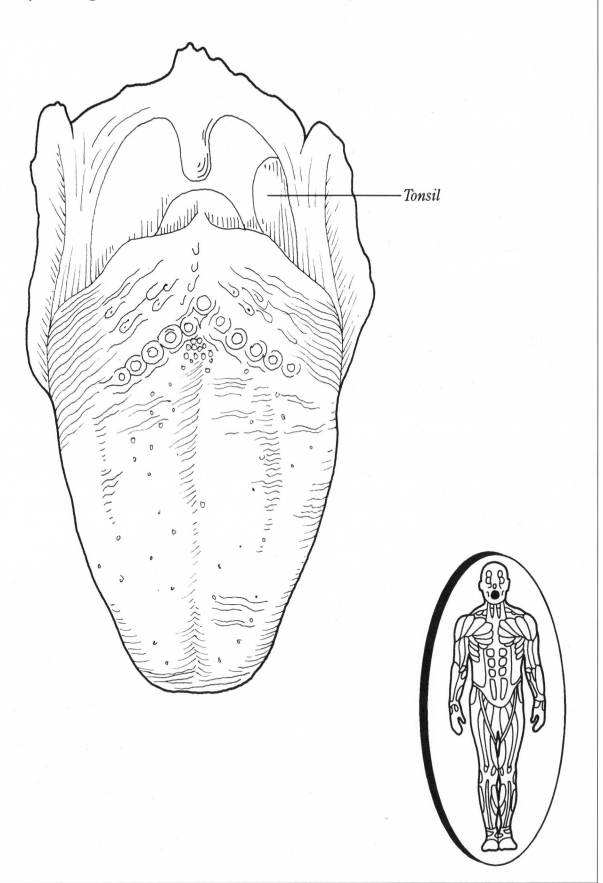

Tonsil

Down the Hatch

Your *voice box* is also known as the *larynx* (LAR-inks). It is made of hard tissue called cartilage. The upper cartilage is called the *epiglottis* (ep-ih-GLOT-tis). The lower cartilage is called the *cricoid* cartilage.

At the back of your throat is a part of your body called the *pharynx* (FAR-inks). The pharynx is the place where both food and air pass to get to the tubes that will carry air to the lungs or food to the stomach. If food goes down the wrong tube, you might choke. The epiglottis keeps food from going down the wrong way.

When you swallow, muscles attached to the larynx pull it up against the epiglottis. The epiglottis acts like a lid and shuts the opening in the larynx that leads to the *trachea* (windpipe). You can feel this happen if you feel your larynx or ''Adam's apple'' while swallowing slowly.

The cartilage that forms the Adam's apple is called the *thyroid cartilage*. It acts as a shield to guard the delicate *vocal cords*. The *arytenoid* (ar-EE-te-noyd) cartilages are attached by many small muscles to the thyroid cartilage, cricoid cartilage, and the vocal cords. These muscles move your vocal cords. When your vocal cords move, your voice can become either higher or deeper.

These cartilages grow and change shape as a child becomes a teenager. This causes the voice to change and is especially noticeable in boys.

The drawing on the left is a view of the larynx, or voice box, as seen from behind. The larynx is made of separate pieces of cartilage. The picture on the right shows a side view of the mouth and throat, showing the structures that let you breathe and swallow.

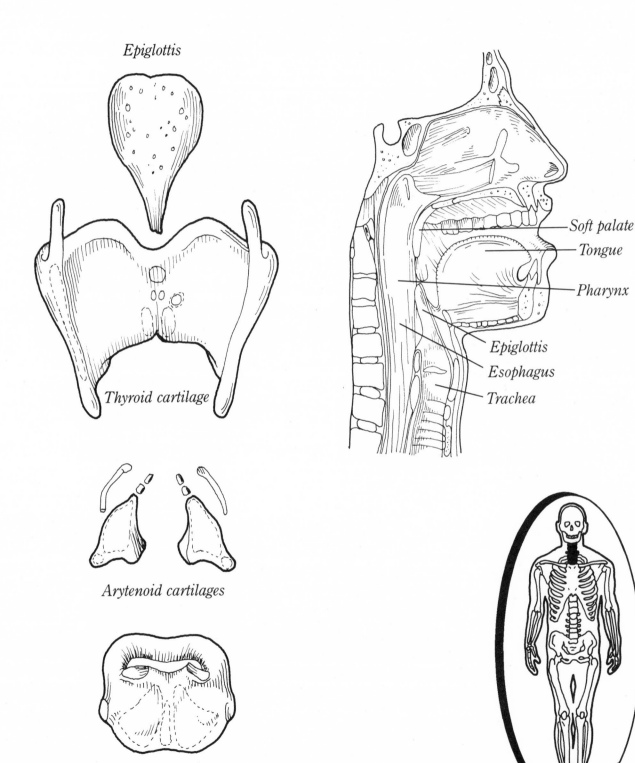

Epiglottis

Thyroid cartilage

Arytenoid cartilages

Cricoid cartilage

Soft palate

Tongue

Pharynx

Epiglottis

Esophagus

Trachea

Your Suit of Armor

Your vital organs are well protected. Just as your brain is in its hard, bony cranium, the organs of your chest are protected by the *sternum, ribs*, and *vertebrae* (VER-tih-bray). Together, these structures make up the *thorax*, or *rib cage*. Your chest isn't made of solid bone because it must move to allow you to breathe.

The *sternum*, or *breastbone*, protects the center of your chest, where your heart and lungs are. The first seven pairs of ribs (there are twelve pairs in all) attach directly to the sternum by their own joints. They are called the *true ribs*. Three pairs of ribs attach only to the sternum through another rib. They are called the *false ribs*. The eleventh and twelfth ribs aren't attached to the sternum at all. They're called the *floating ribs*. All the ribs are attached to your back.

Many muscles are attached to your rib cage. The *intercostal muscles* are found between each rib. They contract to help you breathe properly. You can see them working as your chest rises and falls with every breath. On each side of your chest you have a pair of *pectoral muscles*. You can feel the larger pectoral muscle on either side of your sternum. The smaller one is found beneath the larger one. The pectoral muscles help move your upper arm and shoulder joints.

Your organs are protected by your sternum and the ribs in front (top picture), and by your spinal cord in back. The bottom drawing shows the segments of backbone in your chest region. They're called thoracic vertebrae.

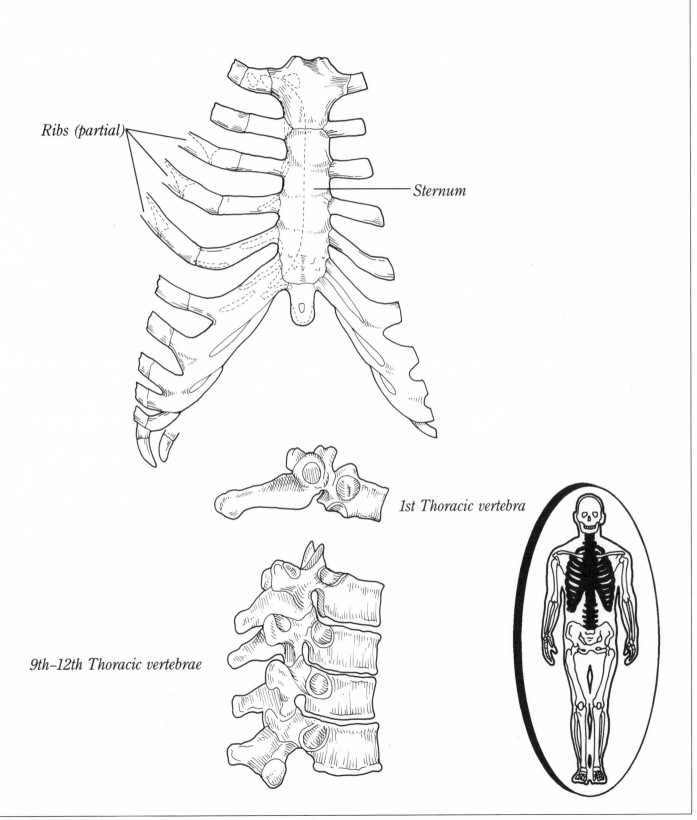

Ribs (partial)

Sternum

1st Thoracic vertebra

9th–12th Thoracic vertebrae

Shouldering The Burden

When you throw a ball, swim, or just raise your arm, you usually don't give it much thought. However, a complex arrangement of bones, joints, and muscles acts together so you can move and rotate your shoulders.

The long bone of your upper arm is called the *humerus* (HEU-mur-us). It's attached by muscles and ligaments, which are tough bands of connective tissue, to two other bones, the *clavicle* (KLA-vik-ul) at the top of your chest, and the *scapula* (SKAP-u-luh) behind your chest.

The top drawing shows the shoulder as seen from the front. The clavicle, or collarbone, attaches to the sternum and helps strengthen your shoulder. It's easy to feel your clavicle on either side and slightly in front of your neck. If your clavicle broke, you wouldn't be able to raise your arm above your shoulder. Breaks or fractures of the clavicle are common, but they usually heal quickly and completely.

This drawing also shows the smaller pectoral muscle, the *pectoralis minor* (PEK-tor-AL-is MY-ner). It's normally hidden beneath the larger pectoral muscle, the *pectoralis major.*

The bottom drawing shows the scapula, or shoulder blade. Sticking out of the scapula are two bony points, the *coracoid* (KOR-uh-koyd) and the *acromion* (ah-KRO-mee-on). Some of the muscles and ligaments that move and strengthen the joint are attached to these bony areas.

The top figure shows the bones of the shoulder area. Note the fractured clavicle. Below is the left shoulder blade, or scapula, as seen from behind.

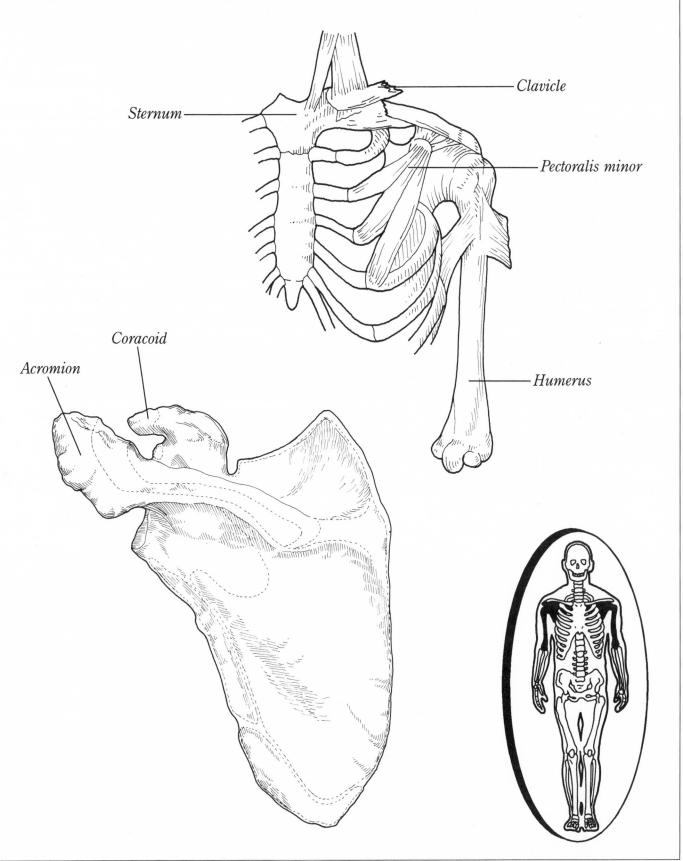

Sternum

Clavicle

Pectoralis minor

Coracoid

Acromion

Humerus

Be Hip!

When you walk, your weight is supported by your *pelvis*, or *hips*. The hip bone is actually made of three bones which have grown together. They are called the *ilium*, the *ischium*, and the *pubis* (IL-e-um, ISS-ke-um, and PYU-bis).

You can feel the ilium at the top of your hips. The pubis is located right above your groin, and you sit on the bony tips of the ischium. These bones provide a strong and stable platform for supporting the organs of your abdomen and pelvis, and they also provide a place for the muscles of your hips and thighs to attach.

The top drawing shows the *femur* (FEE-mur), or *thighbone*, forming a joint with the hip. This arrangement allows the femur to move like the antenna on your radio.

At the back of the hip, you can see the *sacrum* (SAY-krum) attached to the ilium. The sacrum is at the base of your spine and is made of five fused vertebrae. Spinal nerves leave the spinal column through the holes in the sacrum. The *lumbosacral joint* (LUM-bo-SA-krul) is part of the spine and the inflexible sacrum. This area can be strained easily and causing pain in the lower back or even a damaged or ''slipped'' disc.

Your hips are also known as your pelvis. In the top drawing you can see its wide bones, which help support your body's weight. The bottom drawing shows the sacrum, located at the back of the pelvis. The holes allow nerves from the spine to pass through.

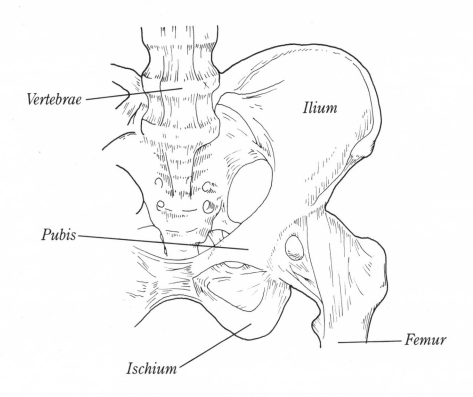

Vertebrae

Ilium

Pubis

Ischium

Femur

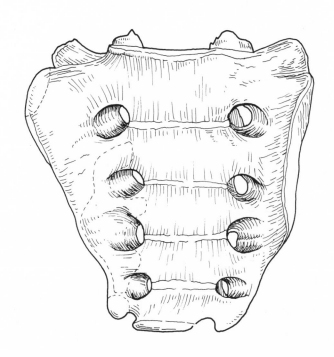

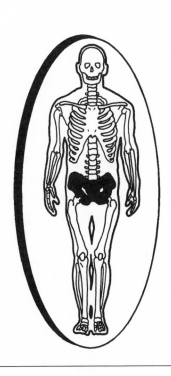

Shake a Leg

The bony *pelvis* is shaped differently in men and women. The male pelvis is heavier and thicker, while the female pelvis is wider and shallower, with a wider *pelvic inlet*. The shape of a woman's pelvis makes it easier to carry and deliver babies than if her pelvis were shaped like a man's.

The *coccyx* (kok-siks) is a group of four small bones at the lower end of the sacrum. Together, they are commonly called the *tailbone*. They are the last part of the vertebral column.

The *acetabulum* (AS-eh-TAB-yuh-lum) is the socket for the end of the *femur,* or *thighbone*. Since the end of the femur is round, this joint is called a *ball and socket joint*. Ball and socket joints allow a wide range of motion, letting you run, dance, and play sports.

The *sacroiliac joint* (SA-kro-IL-ee-ak) is located between the *sacrum* and the *ilium*. When you fall or jump from a height and land on your feet, this joint transfers most of your weight to your hipbones. This prevents injuries to the spinal column.

LEFTOVERS

Did you know that you have a tail? That's what your coccyx is—the remains of a tail!

A tail may have been useful to ancestors of humans who lived millions of years ago, but we have no need for a tail these days. Your coccyx is simply a useless part that's been left over.

Any organ that used to have a function but doesn't anymore is called a vestigial organ. Your tail is vestigial; your dog's tail isn't.

Many times, an organ will be called vestigial because we don't know what it does. The appendix and the tonsils were once thought to be vestigial. Now they're considered to be useful in helping our bodies fight disease.

A man's pelvis, shown in the top drawing, is heavier and thicker than a woman's. A woman's pelvis, shown in the bottom drawing, has a wider pelvic inlet to ease childbearing.

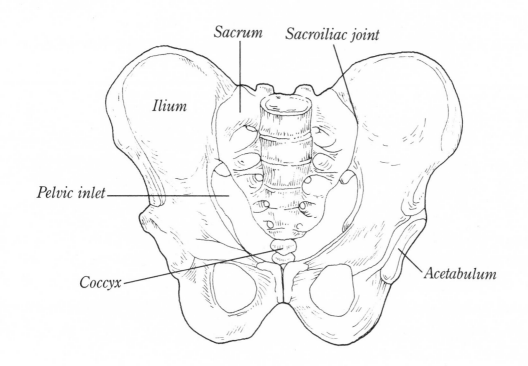

Sacrum   Sacroiliac joint

Ilium

Pelvic inlet

Coccyx

Acetabulum

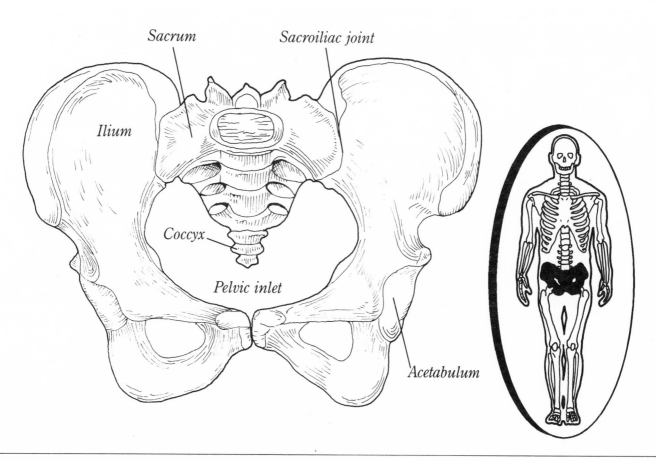

Sacrum   Sacroiliac joint

Ilium

Coccyx

Pelvic inlet

Acetabulum

Chest of Wonders

The section of your body between your neck and your waist is called the *thorax*, or chest. The section below your waist and above your legs is called your *abdomen*, or belly. This drawing shows the organs of the thorax and some of the organs of the abdomen.

Just above the thorax, in the neck, you can see the *thyroid cartilage* of the *larynx* (also called your voice box or Adam's apple), the *trachea*, or *windpipe*, and the paired *internal jugular veins*. The larynx and trachea allow air to travel to and from the *lungs* in the thorax.

The jugular veins drain blood from the head and neck to the *superior vena cava*. This very large vein brings blood from the upper part of your body and arms to the right side of your heart.

The thorax is protected by the *ribs*, *sternum* (breastbone), and *vertebral column*. The *heart* and *lungs* are in the cavity formed by these bones.

Your heart has four chambers that receive and pump blood. The right side of your heart receives blood from your body and pumps it to your lungs. The left side of your heart receives blood from your lungs and pumps it back to your body.

Your lungs have two major tasks. They supply oxygen to your blood so your cells can produce energy to keep your body running. They also remove carbon dioxide, the waste gas, from your blood. Carbon dioxide is made by your cells when they produce energy.

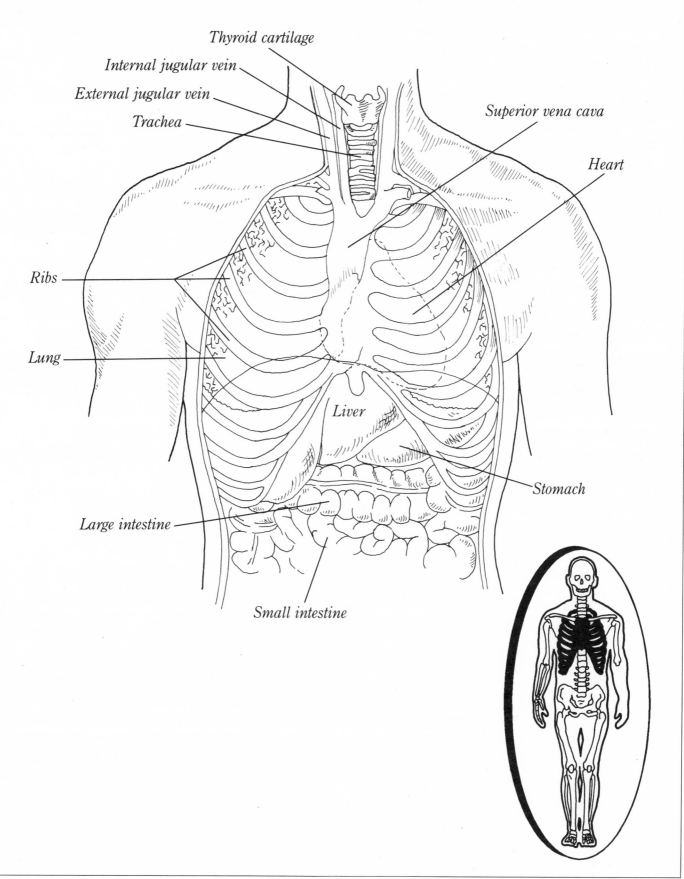

A view of the front of the chest, or thorax, showing some of the major organs.

Thyroid cartilage

Internal jugular vein

External jugular vein

Trachea

Superior vena cava

Heart

Ribs

Lung

Liver

Stomach

Large intestine

Small intestine

Taking a Detour

A blocked artery is a very dangerous thing.

When an artery becomes clogged, blood leaving the heart can't reach the rest of the body. Luckily, your body has backup systems for the movement of blood called *collateral circulation*.

Here's one system of arteries that bypasses the aorta, the largest artery of your body. This system is made of three arteries: the *internal thoracic* (thaw-RASS-ik) *artery*, and the *superior* and *inferior epigastric* (ep-ih-GASS-trik) *arteries*.

The internal thoracic artery is connected to the top of the aorta through the *subclavian artery*. It runs underneath your ribs just to the outside of your sternum. Its branches run beneath each rib, and are called the *intercostal arteries*.

When the internal thoracic artery passes to the front of the abdomen it is called the superior epigastric artery. Its branches join the inferior epigastric artery. The inferior epigastric artery is a branch of the *external iliac artery*, which connects to the bottom of the aorta. This is your body's backup system: a loop of arteries that may keep blood flowing in case one or more of these arteries become blocked.

Near the area where the inferior epigastric artery branches, is the *inguinal ligament*. This ligament marks the boundary between your abdomen and your leg. The inguinal region is weak in certain spots because of a tunnel, called the *inguinal canal*, through the abdominal muscles. Sometimes, when a person strains to lift a heavy object, this area weakens further.

The result is called an *inguinal hernia*. In an inguinal hernia, a part of your bowel, usually the small intestine, bulges through the abdominal wall. It's a serious condition that requires surgery.

The left side of the chest, seen from the inside, showing the ribs and some of the major arteries.

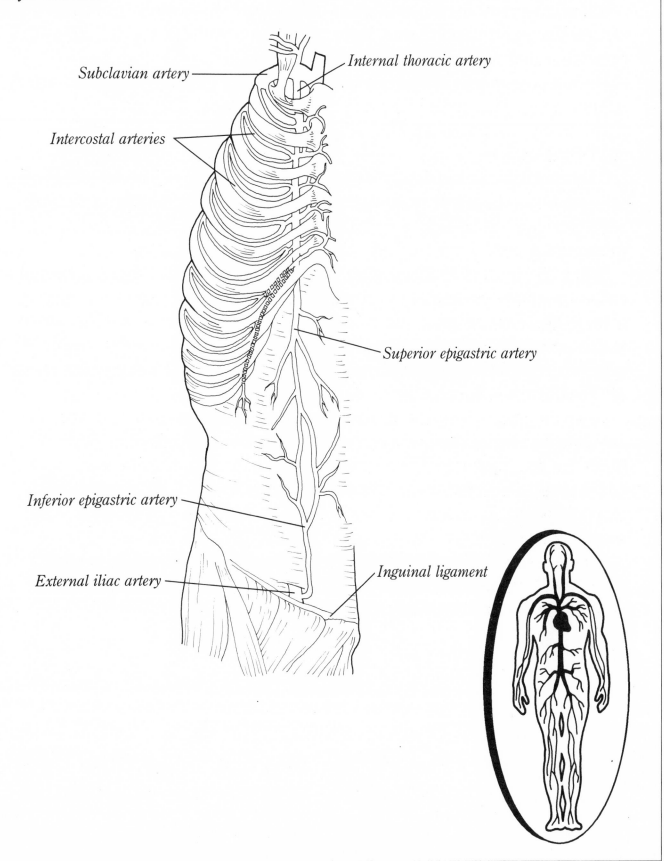

Subclavian artery

Internal thoracic artery

Intercostal arteries

Superior epigastric artery

Inferior epigastric artery

External iliac artery

Inguinal ligament

Take a Deep Breath

The top drawing shows the *heart* and *lungs* as they normally appear.

From this picture you can get an idea of how closely your heart and lungs work together. Two great veins bring blood from the body into the heart. The upper one is called the *superior vena cava* and the lower one is called the *inferior vena cava*. These large veins empty into the upper chamber, or atrium, of the right side of the heart. Blood is pumped from the right side of the heart to the lungs through the *pulmonary arteries*.

At the lungs, carbon dioxide is exchanged for oxygen. Your cells use oxygen to make energy, and they return carbon dioxide as waste. The red color of blood is caused by *hemoglobin*, which carries the oxygen. The oxygen-rich blood returns to the left side of the heart through the pulmonary veins. (You can't see these veins in this drawing.)

The left side of the heart pumps blood through the largest artery in the body, the *aorta* (AY-or-tuh). The aorta is about an inch wide. It curves down the left side of your spine into your abdomen. Your abdomen is the area below your waist and above your legs.

Your *lungs* are the organs of respiration, or breathing. You breathe about twelve times a minute, drawing more than 450 million breaths in a lifetime! Air, which is full of oxygen, enters your nose or mouth and passes through the pharynx and larynx (voice box). It then enters the *trachea* (TRAY-key-uh). The trachea branches into the two main tubes, called *bronchi* (BRON-ky). Each bronchus (singular for bronchi) divides into smaller and smaller tubes. The smallest tubes are called *bronchioles* (BRON-key-ols), and they open into the microscopic *alveoli* (al-VEE-o-li), or *air sacs*.

It is in the tiny alveoli that gas is transferred from your lungs to your blood, and from your blood to your lungs. With every breath, you take in oxygen and breathe out carbon dioxide.

Your heart and lungs work closely together to keep blood and oxygen flowing throughout your body. At top is a front view of these vital organs. At bottom are shown the "windpipes" that air travels through on the way to your lungs—the trachea and bronchi.

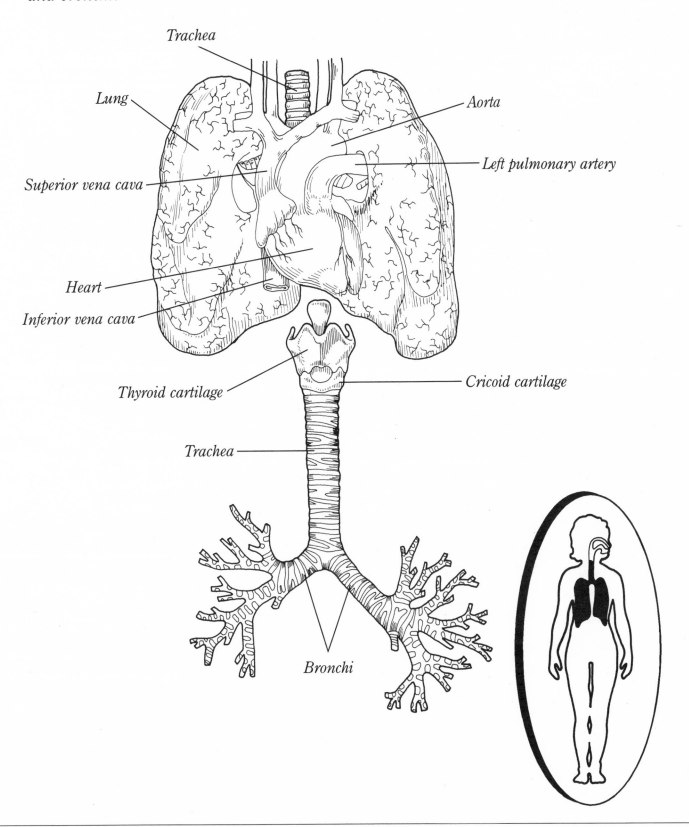

Trachea

Lung

Aorta

Superior vena cava

Left pulmonary artery

Heart

Inferior vena cava

Thyroid cartilage

Cricoid cartilage

Trachea

Bronchi

Have a Heart

Your *heart* is a remarkable muscle. It contracts day and night, more than 2½ billion times during a lifetime without stopping. A healthy heart can pump more than 10 gallons of blood a minute through 60,000 miles of blood vessels! A round trip through this circulatory system takes less than half a minute.

Your heart has four chambers. The two top chambers are called *atria* (AY-tree-uh); the bottom chambers, which are larger than the atria, are called *ventricles*.

The *right atrium* (singular for atria) gets blood from the body through the *superior* and *inferior vena cava*. The right atrium then pumps blood through a valve into the *right ventricle*, which pumps the blood through another valve into the *pulmonary artery*. The pulmonary artery divides into two arteries and delivers blood to the lungs. In the lungs, the blood picks up oxygen and gets rid of waste gases.

From the lungs, blood travels through the *pulmonary veins* to the *left atrium*. The left atrium pumps the blood through a valve into the *left ventricle*, which pumps the blood through another valve into the *aorta*. The aorta delivers the blood to the rest of the body.

When you listen to your heartbeat through a stethoscope you can hear a "lub-dub" sound. The first sound is your atria pumping blood into your ventricles. The second sound is the pumping of your ventricles. By placing the stethoscope at different places on your chest, a doctor can hear the sounds of your heart valves as well.

The heart has its own blood supply from a set of arteries called the *coronary arteries*. These arteries may become narrower as a person becomes older, especially if he or she eats fatty food and doesn't exercise. Fats can build up on the walls of coronary arteries and may eventually close them off completely.

This may also happen in other arteries of the body. When an artery becomes blocked, oxygen and other nutrients can't reach the cells and they may die. When this happens in the heart, it's called a heart attack.

Your heart is a pumping station with four chambers that receive and pump out blood. At top is a rear view of the chambers on the left side of the heart. At bottom is a front view of the smaller chambers on the heart's right side.

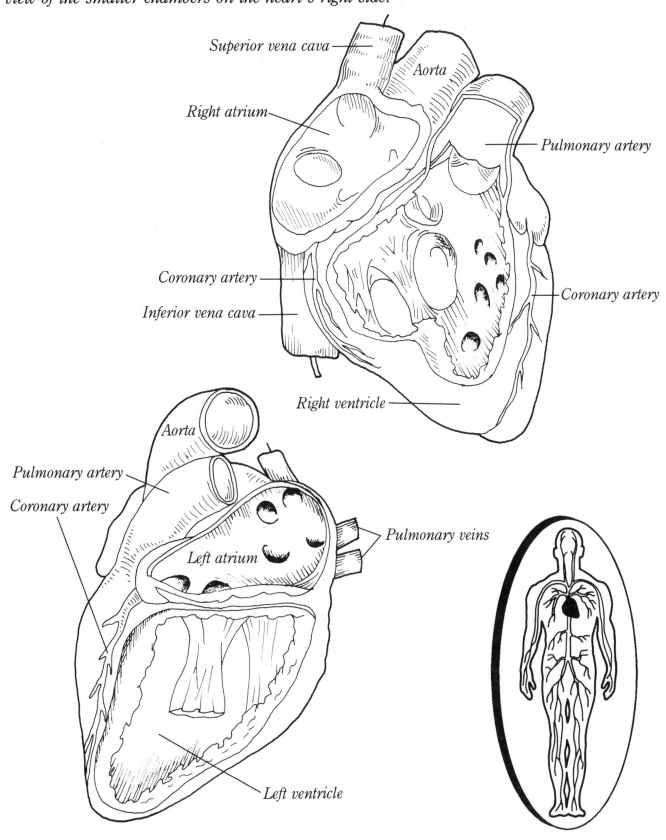

Superior vena cava

Aorta

Right atrium

Pulmonary artery

Coronary artery

Inferior vena cava

Coronary artery

Right ventricle

Aorta

Pulmonary artery

Coronary artery

Left atrium

Pulmonary veins

Left ventricle

A Look at Your Thorax

This drawing shows a view of the *thorax* (chest) from behind without the spine. Let's work our way down to see how some of the major blood vessels, nerves, and organs connect with one another.

In the upper part of the drawing you can see the *pharynx* and the *esophagus*, or *gullet*, which arises from it. The *larynx* (voice box) is also connected to the pharynx, but you can't see it in this picture.

From the larynx, the *trachea* travels into the chest cavity and branches into the *bronchi*, which enter the lungs. The *internal jugular veins* travel from the top of the lungs to the neck. The *pulmonary veins* go from the center of the lungs to the left atrium of the heart.

From this drawing you can see how important the aorta is. It supplies blood from the heart to the head and neck through the *common carotid arteries*. It also carries blood to the shoulders and arms through the *subclavian arteries*.

On the right, you can see the right *vagus nerve*. The vagus nerves are some of the longest nerves in your body. Their most important job is to control the muscles of the pharynx, larynx, and vocal cord, allowing you to speak. They're also responsible for slowing down the heart and supplying sensation to the abdominal organs, such as your stomach and intestines.

The muscle at the bottom of the lungs is the *diaphragm* (DY-uh-fram). It separates the chest cavity above from the abdominal cavity below. Working with the *intercostal muscles* between the ribs, the diaphragm helps you to breathe when it contracts.

WHAT CAUSES HICCUPS?

Hic!—I can't stop—Hic!—hiccuping!

My—Hic!—stomach and intestines are under my—Hic! —diaphragm, and they sometimes tickle it. Hic! Maybe it's because I—Hic!— ate too fast. Sometimes holding my breath—Hic!— helps.

Ahh! That's better.

Normally, your diaphragm helps you breathe. But sometimes it gets tickled, causing hiccups. Holding your breath or sipping water can stop your diaphragm from rubbing against your stomach. And that puts an end to your hiccups!

The organs in your neck and chest, as seen from behind. (The spine was left out of this drawing to give you a better view.) Notice how much room your lungs take up.

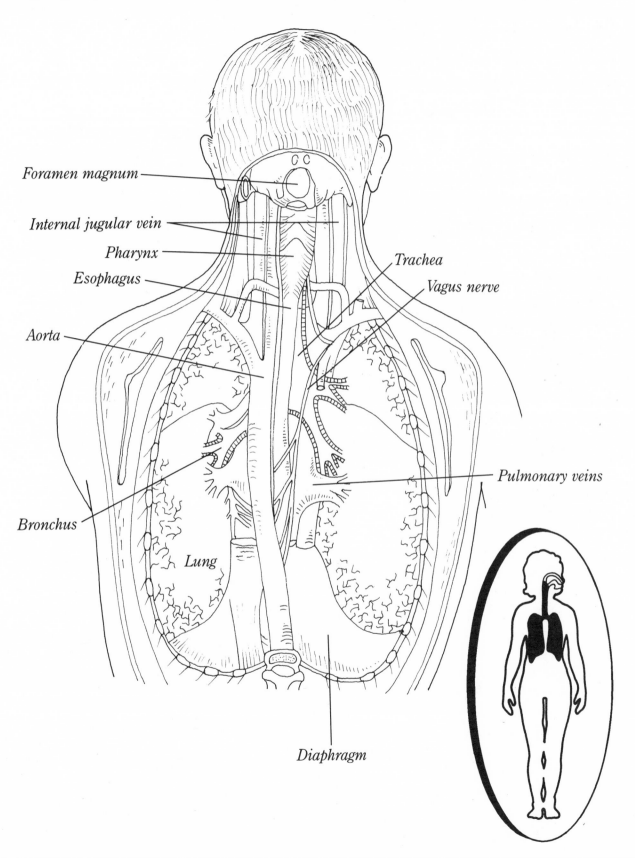

Foramen magnum

Internal jugular vein

Pharynx

Esophagus

Aorta

Bronchus

Lung

Trachea

Vagus nerve

Pulmonary veins

Diaphragm

Food into Energy

Here's a peek at some of the major organs of the *abdomen*, or *belly*.

The muscles of the abdominal wall and a fatty apron-like covering over the intestines have been left out of the drawing. You can clearly see the *liver*, the *stomach*, the *large intestine*, the *small intestine*, and the *bladder*.

Your *liver* is a large, reddish-brown organ that's divided into several parts, called lobes. The largest lobe is partly covered by the right side of your rib cage. Your liver stores sugar, produces important proteins, and filters harmful chemicals from your blood.

Just beneath your liver is your stomach. It's a large sac where the protein in your food starts to get digested. When food is digested, it's converted into chemicals that your body needs for energy, growth, and repair.

From your stomach, food passes through several parts of your small intestine where digestion continues and food is absorbed into the blood.

Your large intestine, or *colon*, absorbs most of the water that remains after digestion and forms solid feces. Feces is the waste product of the foods you eat.

Your bladder stores the liquid waste produced by the kidneys.

The *common iliac arteries* are the end branches of the aorta in the abdomen. They supply blood to your bladder and other organs in the pelvis as well as to your hips and legs.

The major organs in your abdomen, or belly, help turn food into the chemicals your body needs to grow and heal. Bones of the shoulder, rib cage, and pelvis are also shown.

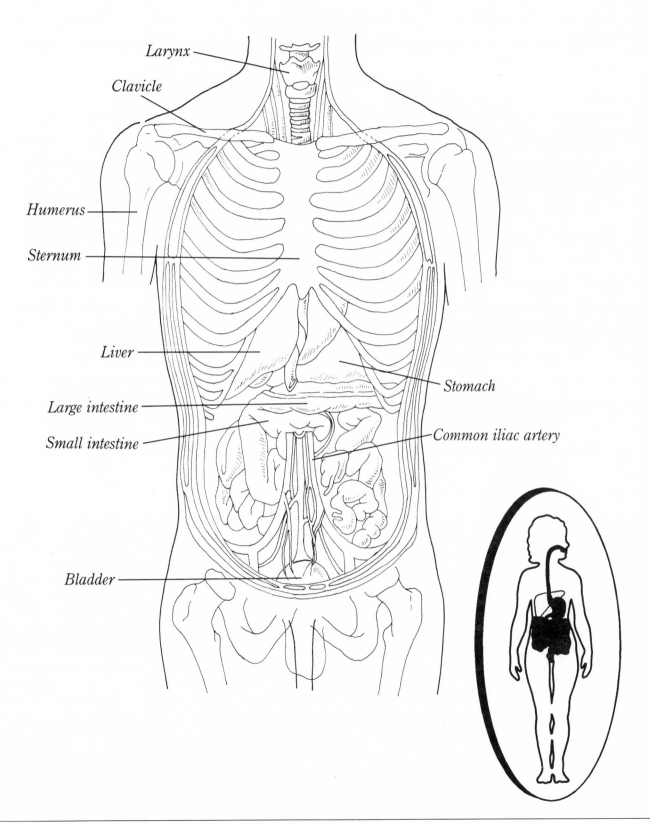

Larynx

Clavicle

Humerus

Sternum

Liver

Large intestine

Small intestine

Bladder

Stomach

Common iliac artery

Digestion is the breakdown of food into nutrients your body can use. It starts in your mouth as you chew your food into small pieces or a thick paste.

When you swallow, food travels into the *pharynx*, down the *esophagus*, and into your *stomach*. Your stomach then secretes powerful acid and an enzyme called *pepsin* to dissolve the food.

Sometimes a little acid may escape from your stomach into your esophagus. This causes the burning sensation in your chest known as heartburn.

You may be wondering why the acid and pepsin in your stomach don't dissolve your stomach along with your dinner. Your stomach is protected from digesting itself by a thick coating of mucus. For some people, this protection isn't enough, and the lining of the stomach does get damaged by the acid. When this happens, the damaged area is called an ulcer, and a doctor's help is needed.

After leaving the stomach, partially digested food enters the first part of the small intestine called the *duodenum* (DU-uh-DEE-num). In the duodenum, the acid loses its strength and digestion continues. Many chemicals are secreted by the *pancreas* (PAN-cree-us) into a tube that connects to the duodenum. This tube, called the *main pancreatic duct*, joins with the *common bile duct* as it enters the duodenum.

The common bile duct delivers *bile* from the *gall bladder* to the duodenum. Bile is a substance that helps digest fats. It's produced in your liver and stored in your gall bladder.

The *spleen* isn't an organ of digestion. It destroys old and worn-out red blood cells. It's found in the upper abdomen under the left side of the lower rib cage. It shares a blood supply with the stomach and pancreas.

Your stomach contains powerful acid to dissolve the food you eat. The stomach is drawn upside down in the top drawing to show you the organs behind it, along with the major arteries. At bottom, a cutaway drawing shows the stomach in its normal position.

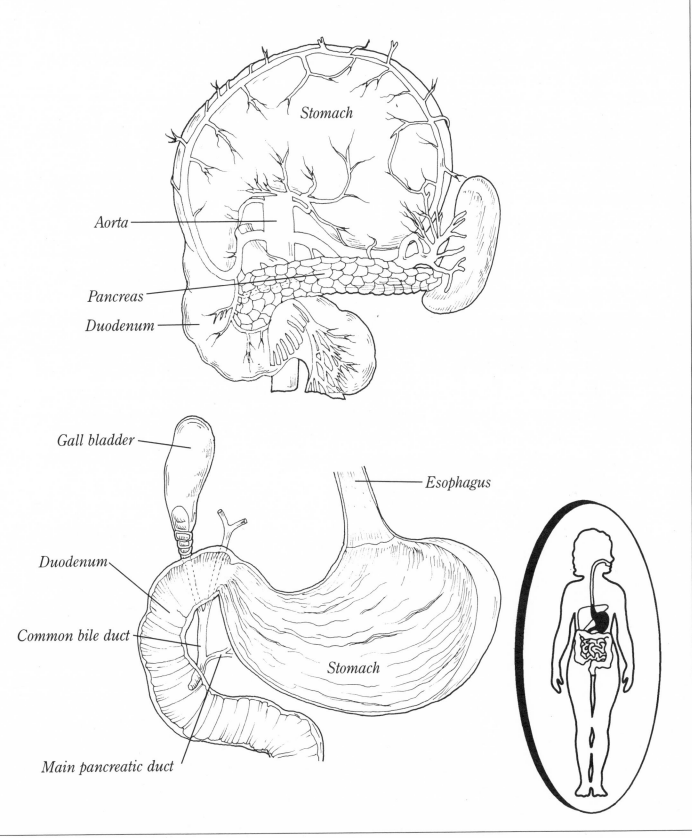

Stomach

Aorta

Pancreas

Duodenum

Gall bladder

Esophagus

Duodenum

Common bile duct

Stomach

Main pancreatic duct

The Factory in Your Belly

In this drawing, part of the *duodenum* and *colon* have been removed to show the *portal system of veins*. The portal system drains blood from the digestive organs into the liver.

The portal system is made of the *superior mesenteric vein*, the *inferior mesenteric vein*, the *splenic vein*, and the *portal vein*.

The superior mesenteric vein drains blood from the *small intestine* and the right side of the *large intestine*, or *colon*. The blood in the left side of the colon is drained by the inferior mesenteric vein. This vein joins with the splenic vein and superior mesenteric vein to form the portal vein.

The splenic vein drains blood from the *spleen*, the *duodenum*, and the *stomach*.

Your *liver* is like a chemical processing factory and has many jobs. Its cells can turn sugar into a starch called *glycogen* and store it for later use. When sugar levels in your blood drop too low, your liver will break down its glycogen and release sugar into the blood for energy.

Liver cells can also take some poisons and drugs out of your blood and change them so they cannot hurt you. Unfortunately, the system is not perfect, and the liver itself may get poisoned. This often happens to people who drink too much alcohol. Alcohol may damage the liver beyond repair and can put a person's life in danger.

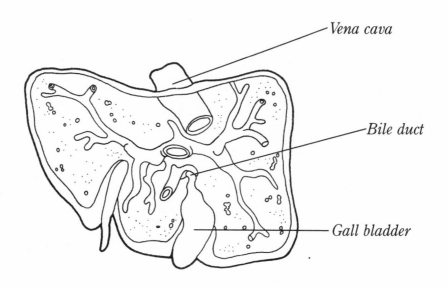

Vena cava

Bile duct

Gall bladder

Your liver is the largest gland in your body.

Blood travels from your intestines to your liver through the portal system of veins. In this picture, the front of the liver has been lifted to show you what's beneath it.

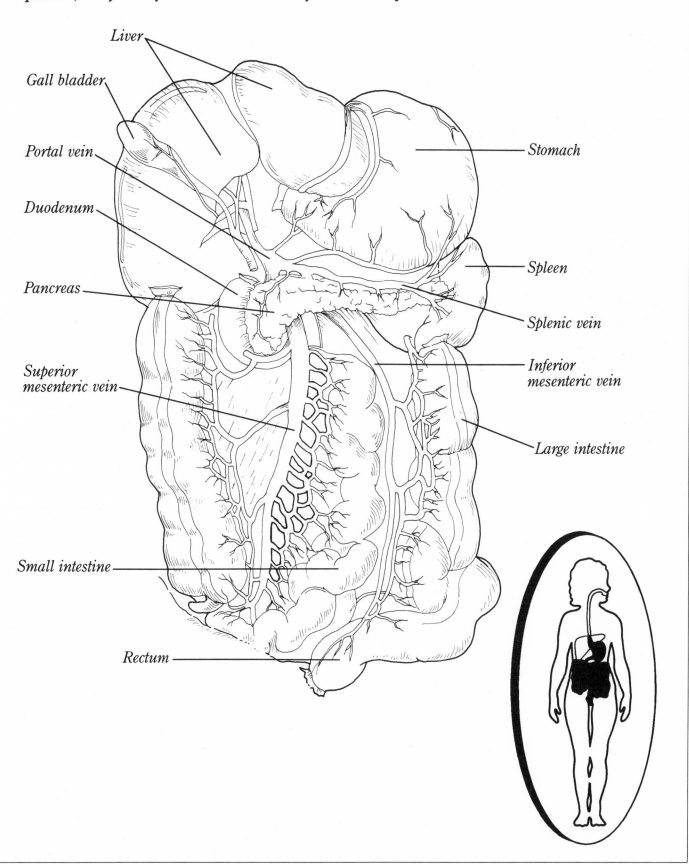

Liver

Gall bladder

Portal vein

Duodenum

Pancreas

Superior
mesenteric vein

Small intestine

Rectum

Stomach

Spleen

Splenic vein

Inferior
mesenteric vein

Large intestine

Passing Through

The *small intestine* is a tube with many loops and coils. If it were stretched out, it would be more than 20 feet long! Most of the food we eat gets digested and absorbed in this organ.

The inside of the small intestine has millions of almost microscopic "fingers" called *villi* which stir and transfer digested nutrients into your blood. The end of the small intestine nearest the *large intestine*, or *colon*, has groups of cells called lymphocytes. These cells help fight any disease-causing bacteria that get into your digestive system.

The walls of the pharynx, esophagus, stomach, small intestine, colon, and rectum are lined with muscle. The muscle pushes food along the digestive system by a wavelike contraction called *peristalsis* (per-ih-STAL-sis). These waves are coordinated to keep food moving in one direction.

Pushed by peristalsis, the food reaches a valve at the end of your small intestine. This valve allows the undigested food and water to enter the colon. The colon absorbs most of the water from the food passing through. Bacteria that live in the colon break down much of whatever food remains. The *rectum* helps expel the remaining waste, called feces. It's the last part of the digestive system.

WHAT'S YOUR TUMMY TELLING YOU?

Does your stomach growl when you're hungry? It isn't really your stomach that's making all the noise—it's your large intestine.

Breakfast goes to your large intestine at about the time your stomach is getting ready for lunch. You feel hungry because your stomach is empty. When you hear all the growling you might think it's your stomach calling for lunch, but actually, the noise is caused by the walls of your large intestine pushing your breakfast through.

Most of the food you eat gets digested in the small intestine (the small loops shown at left). Then it passes into the large intestine, which circles around the abdomen and becomes the rectum.

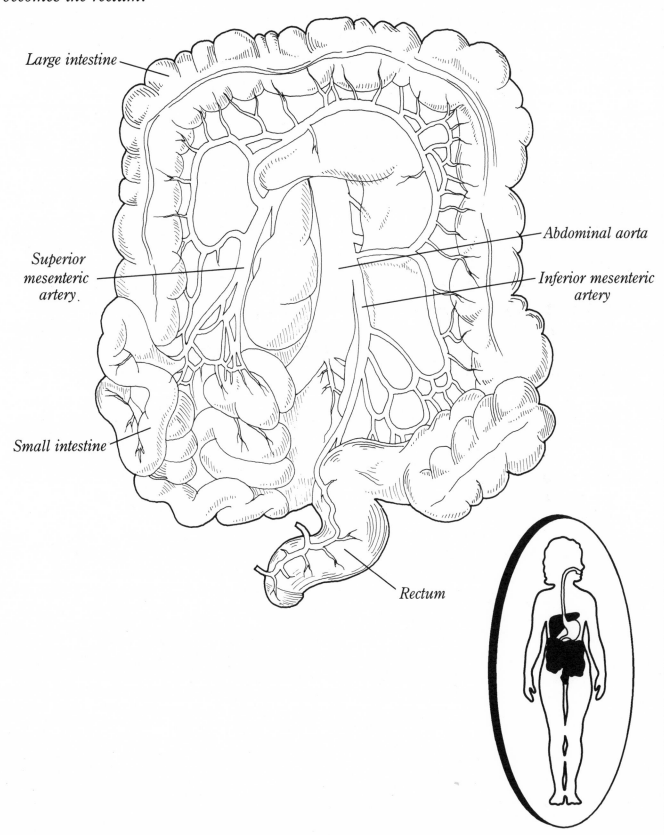

Large intestine

Abdominal aorta

Superior
mesenteric
artery

Inferior mesenteric
artery

Small intestine

Rectum

A Lot of Guts

From this drawing you can see the positions of the abdominal and pelvic organs in the *abdominal cavity*.

Your *liver* lies directly under your *diaphragm* in the upper part of your abdomen. Your *stomach* has an apron-like sheet of tissue called the *greater omentum*, which drops over your intestines and is connected to the *transverse colon*. If a hole, or ulcer, develops in your intestine or stomach, the omentum may help repair it.

The *pancreas* and *duodenum* are at the rear wall of your abdominal cavity. The rest of the *small intestine* and the *colon* are attached to the rear wall by a tissue called the *mesentery*. Blood vessels, lymph vessels, and nerves travel to these organs along the mesentery.

The organs of your pelvis are in the abdominal cavity between the sacrum and pubis bone. These organs are the *rectum*, which expels the solid waste from the colon, and the *bladder*, which collects the liquid waste produced by the kidneys.

Females have additional organs between their rectum and bladder. They are the female reproductive organs, the *uterus* and a pair of *ovaries*. These organs make it possible for women to have babies.

This side view of a woman shows the major organs of digestion as well as the spine and the reproductive organs.

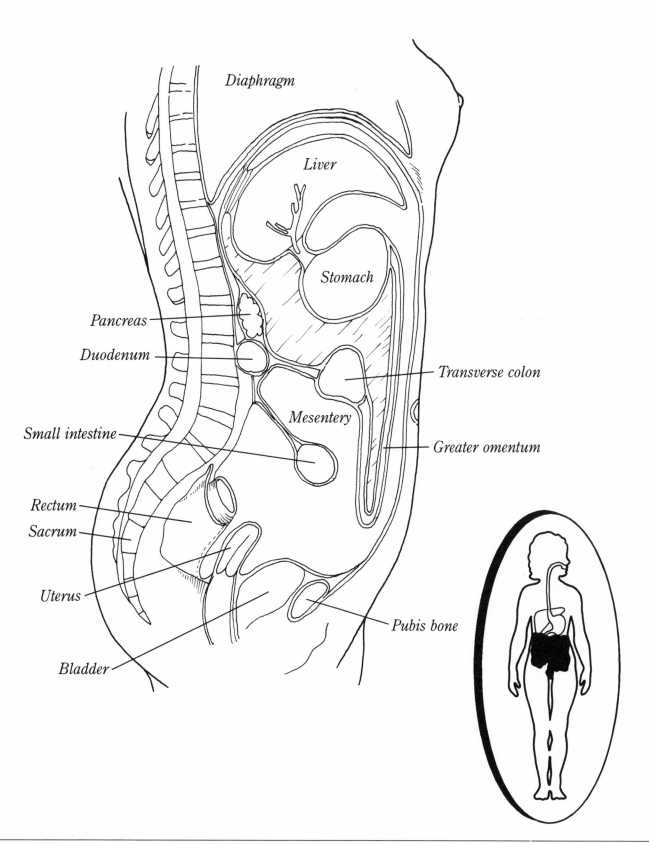

Diaphragm

Liver

Stomach

Pancreas

Duodenum

Transverse colon

Mesentery

Small intestine

Greater omentum

Rectum

Sacrum

Uterus

Pubis bone

Bladder

Automatic Pilot

Your heart rate, breathing rate, blood pressure, and the way your pupils adjust to changing light are all *involuntary functions*, which means you have no control over them. These and similar functions are under the control of your autonomic nervous system.

The autonomic nervous system is divided into two parts which act opposite one another. One part is called the *parasympathetic* (PAH-ruh-sim-puh-THEH-tik) *nervous system*. The other is the *sympathetic* (sim-puh-THEH-tik) *nervous system*.

Your parasympathetic nervous system is made of the paired *vagus nerves*, parts of other nerves in your brain, and some spinal nerves from your pelvic area. The parasympathetic nervous system brings your body to a resting state.

For example, this system slows your heart and breathing rate, but it also increases the wavelike motions of your digestive system when you eat.

On the other hand, the sympathetic nervous system does just the opposite. This system increases your heart rate, quickens your breathing rate, and prepares your body to fight or flee when you're excited or scared.

The sympathetic nervous system is made of the *sympathetic chain* and ganglia. A *ganglion* (singular for ganglia) is a group of nerve cells outside the central nervous system. When ganglia and nerve cells are bunched together, they're called a *nerve plexus*.

You may have heard someone talking about the "solar plexus," also called the *celiac plexus*. It's located behind the stomach, directly beneath your sternum. Try pushing gently on it—you'll feel how sensitive it is.

Boxers have to be careful to guard their solar plexus, or else they can get "the wind knocked out of them." The celiac plexus helps regulate the stomach, gall bladder, liver, spleen, pancreas, and duodenum.

Here are the main branches of the sympathetic nervous system. These control your pulse rate, breathing, blood pressure, and many other bodily functions you never even think about.

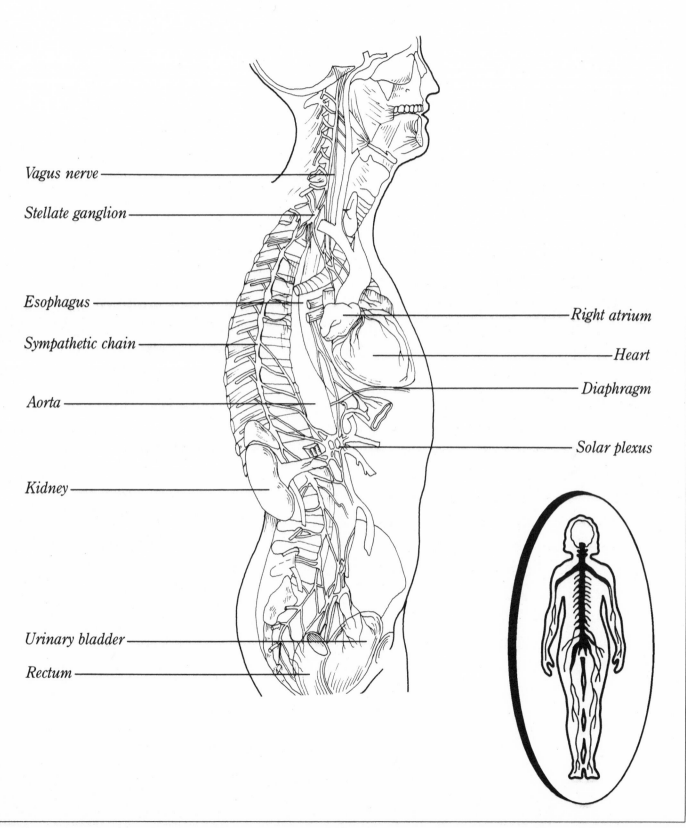

Vagus nerve

Stellate ganglion

Esophagus

Sympathetic chain

Aorta

Kidney

Urinary bladder

Rectum

Right atrium

Heart

Diaphragm

Solar plexus

Chemical Messengers

Your *kidneys* lie at the very back of your abdominal wall, near your 12th rib. They're about an inch from both sides of your spine. The kidneys act as very selective filters of your blood. They remove extra water, salts, and other wastes, which leave your body as *urine*.

Urine travels from the kidneys through the *ureters*, and is collected in the bladder. When your bladder gets full, you know it's time to head for the bathroom.

Just above the kidneys are the *adrenal* (uh-DREE-nul) *glands*. The adrenal glands are part of the *endocrine gland system* and they secrete several hormones, including some steroid hormones. Hormones are chemical messengers that turn body functions on or off.

Steroid hormones can do a lot. They control the salt and water balance in your body, control swelling, help repair damaged tissues, and help you respond when you're aroused.

Besides adrenal glands, other endocrine glands include the *ovaries* in females and the *testes* in males, which secrete sex hormones; the *thyroid gland* in your neck, which secretes hormones that control how you use energy; and the *pancreas*, which secretes hormones that control sugar levels in your blood.

When your glands don't work properly, serious illnesses often result. For example, when the pancreas fails to produce a hormone called insulin, the level of sugar in the blood can rise dangerously. We call this disorder diabetes.

When your kidneys aren't doing their job of balancing the amount of sugar in your blood, sugar appears in your urine. When you give your doctor a sample of your urine, he or she tests it for sugar and other things that shouldn't be there. That way your doctor knows if your glands and kidneys are doing their job.

If your glands control so many delicate operations in your body, what controls your glands? The answer is another gland! The master gland of the endocrine system is your pituitary gland, located at the base of your brain.

Your kidneys filter wastes from your blood. This is a view of them from behind. The spine has been left out to give you a better view.

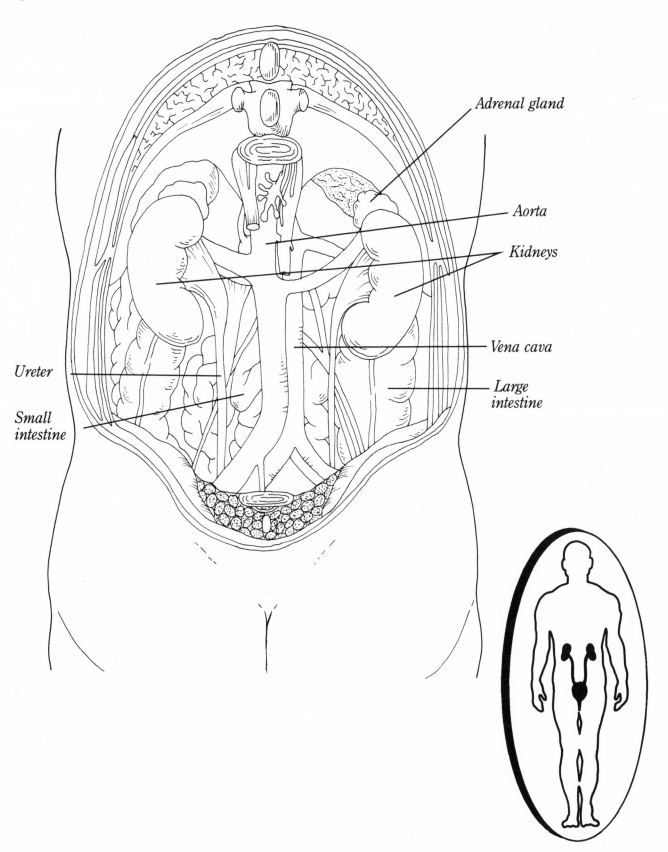

Adrenal gland

Aorta

Kidneys

Vena cava

Large intestine

Ureter

Small intestine

Wave Your Arms

Your shoulder joint and scapula (shoulder blade) can move in many directions. You can lift your arm to the front or side, rotate it forward or backward, and shrug your shoulders. Many muscles are responsible for these actions.

The *deltoid* is the muscle covering your shoulder joint. It can raise your arm to the front, side, and rear, but only to your shoulder's height. In order to raise your arm above your shoulder and rotate it, the muscles of the *rotator cuff tendons*, the *trapezius* (tra-PEE-zee-us), and the *levator scapulae* (leh-VAY-tor SKAP-yu-lay) must work.

Four muscles form the rotator cuff: The *supraspinatus* (SOO-pruh-spin-AH-tis) muscle, located above the scapula's bony spine; the *infraspinatus* muscle, located below the spine; the *teres* (TER-eez) *minor* muscle, located just under the infraspinatus; and the *subscapularis* muscle located underneath the scapula (not shown in the drawing).

All the rotator cuff's muscles attach to the upper arm bone, called the *humerus*, as well as to the scapula. The rotator cuff keeps the humerus in the shoulder joint. Many baseball pitchers suffer rotator cuff injuries because of the strain of throwing baseballs.

You can feel the trapezius and levator scapulae contract when you shrug your shoulders, and you can feel your major and minor rhomboid muscles contract when you pull your shoulders back.

The *latissimus dorsi* (lah-TIS-ih-mus DOR-see) muscle is the wide, fan-shaped muscle of the lower back that attaches to the humerus. You use this muscle when chopping wood, climbing, paddling a canoe, and making similar movements.

The *sternocleidomastoid* (STER-no-KLY-do-MASS-toyd) and *splenius* (SPLEE-nee-us) muscles help move your head, and the *external oblique* muscle helps you bend your body to the side.

Your back and shoulders have many layers of muscles. They allow you a wide range of motion. On the left side is the first layer of muscles. The deeper muscles that lie just below them are shown on the right.

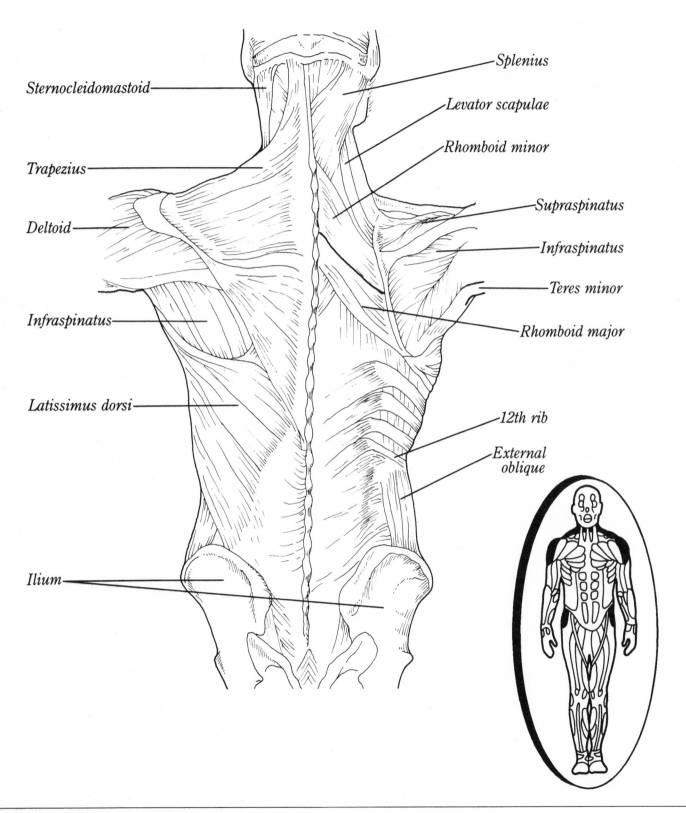

Sternocleidomastoid

Trapezius

Deltoid

Infraspinatus

Latissimus dorsi

Ilium

Splenius

Levator scapulae

Rhomboid minor

Supraspinatus

Infraspinatus

Teres minor

Rhomboid major

12th rib

External oblique

Deep in Your Bones

The *humerus* is the largest bone of your arm and is part of your shoulder and elbow joints. The *femur* is the largest bone of your leg and is part of your hip and knee joints. The femur is also the largest bone in your body.

The shape of the femur and the humerus is similar because the bones have similar jobs. Their rounded heads allow them to move freely in their joints so that you can wave your arms and march around.

Muscles are attached to the outside of these bones. They are indicated by the dotted lines. Inside the bones is tissue called bone marrow.

Bones contain either yellow marrow or red marrow. Yellow marrow is fatty tissue. Red bone marrow contains the cells that make your red and white blood cells. Your femur and humerus contain red marrow.

Doctors perform marrow transplants on people who have a blood disease called leukemia. A person with leukemia produces too many white blood cells.

The doctor first destroys the marrow cells of the leukemia patient, killing the diseased cells. Then he or she uses a syringe to remove some red marrow cells from the hip bone of a healthy donor. Next, the doctor injects the healthy cells into the leukemia patient.

The donated cells go to the patient's bone marrow and start producing healthy cells.

The largest bone in your body is your upper leg bone, the femur, shown on the left. The bone on the right is the upper arm bone, called the humerus. Notice how similar they look.

Femur   **Humerus**

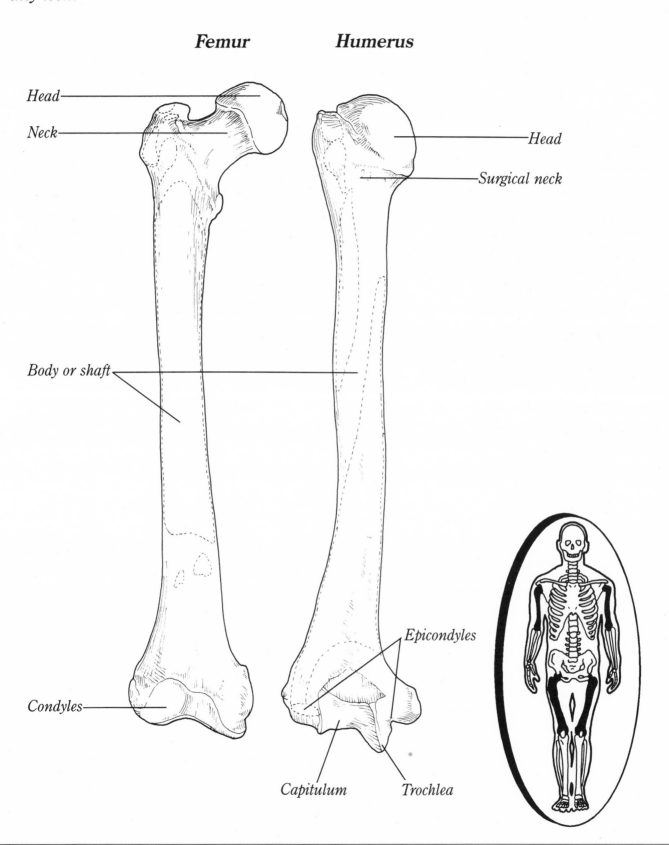

Head

Neck

Head

Surgical neck

Body or shaft

Epicondyles

Condyles

Capitulum   Trochlea

Your forearm is made up of two bones, the *radius* and the *ulna*. The radius is on the thumb side of your forearm, and the ulna is on the pinky side.

The point of your elbow is formed by the *olecranon* (o-LEK-ruh-non) of the ulna. That's where the ulna is attached to the upper arm bone, or humerus.

At its other end, the rounded head of the ulna has a bony bump called a *styloid process*. You can feel it at the back of your wrist on the pinky-side of your forearm.

The disc-shaped head of the radius isn't firmly attached to your upper arm bone. This loose connection lets you turn your hand palm-up and palm-down.

The other end of the radius is wider and more fully connected to your wrist bones. You can feel the bony styloid process of the radius on the thumb side of your wrist, just above the fleshy part of your thumb.

Your *radial artery* is just to the inside of this process. If you feel this area with your fingers, you will be able to feel the artery pulse.

These are the two bones of your forearm, as seen from the front. The bone on your thumb side is called the radius. The other bone (as seen with the palm facing up) is the ulna. Ridges show the attachment of muscles that move the bones.

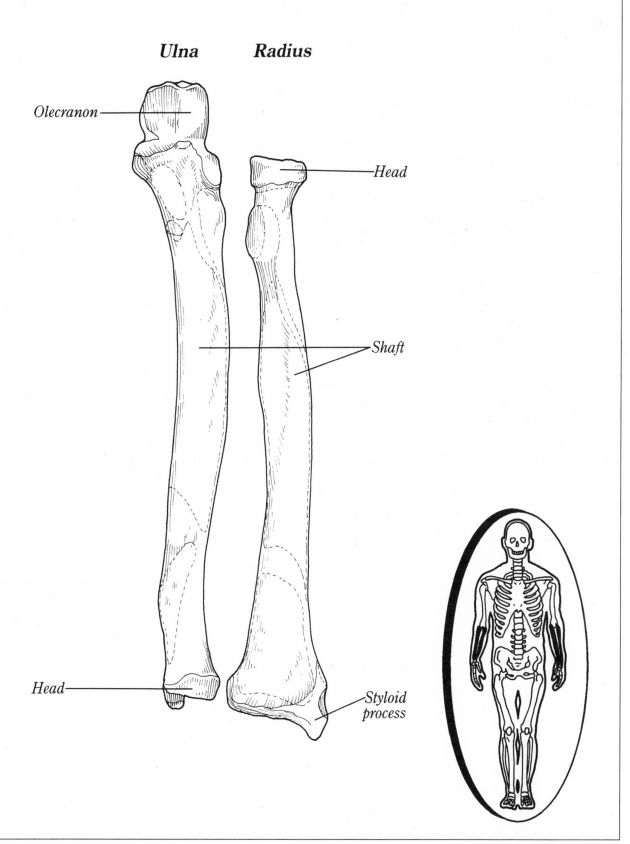

Ulna   **Radius**

Olecranon

Head

Shaft

Head

Styloid process

Your hands are marvelously flexible. Hands that can strongly grip an ax may also delicately play a piano. You have 27 bones between the end of your wrist and the tips of your fingers. All these bones let you move your hands in so many handy ways.

The eight bones in your wrist are called *carpal bones*. The largest carpal bone is the *capitate bone*, but you can see from the picture that it's not very big at all. The other carpal bones are the *trapezium*, *trapezoid*, *hamate*, *pisiform*, *triquetrum*, *lunate*, and *scaphoid*. These bones slide against each other allowing graceful hand movements.

Above your carpal bones are five *metacarpal bones*. These bones have muscles attached to them that allow you to spread your fingers apart and move them closer together.

The bones of your fingers are called *phalanges* (fuh-LAN-jeez). Each finger has three phalanges, and the thumb has two.

The fleshy part of your thumb has four muscles, and the fleshy part of your little finger has three muscles. These muscles allow you to pick up small objects. For example, to grip your coloring pen, pencil, or crayon, you move your thumb opposite to your index finger.

Try getting along without your thumb, even for a short while!

Each of your hands has 27 bones. The eight bones of your wrist are called the carpals. The bones in your palm are the metacarpals. Your finger and thumb bones are the phalanges. The hands in these pictures are drawn with the palms down.

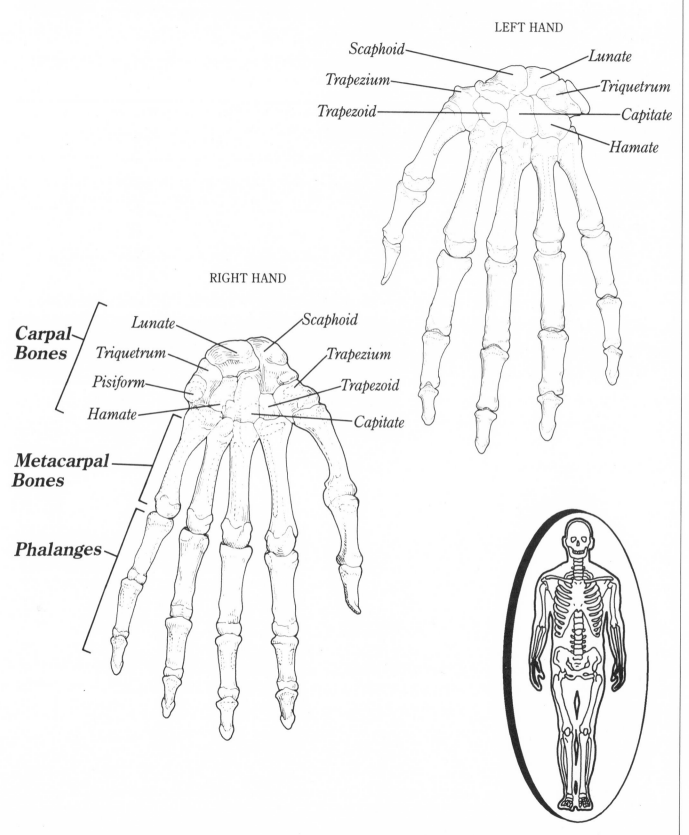

LEFT HAND

Scaphoid

Lunate

Trapezium

Triquetrum

Trapezoid

Capitate

Hamate

RIGHT HAND

Carpal Bones

Lunate

Scaphoid

Triquetrum

Trapezium

Pisiform

Trapezoid

Hamate

Capitate

Metacarpal Bones

Phalanges

Using Your Muscles

To move your arm and bend your elbow, you rely on many muscles.

The *deltoid muscle* allows you to raise your arm to the front, side, and rear. The *pectoralis major* is the large fan-shaped muscle of your upper chest. It pulls your arm toward your body. Your *coracobrachialis* (KOR-ah-ko-BRAY-kee-al-is) muscle helps flex your arm at the shoulder and keeps the joint in place.

The *biceps* is the two-part muscle at the front of your arm. It's the muscle that bulges when you bend your elbow.

The three-part *triceps* is the major muscle at the rear of your arm. When you reach for something on the top shelf, you're using your triceps to extend your arm and your deltoid to raise it at the shoulder. Try to feel all these muscles while moving your arm.

The other major muscle of the upper arm is the *brachialis* (BRAY-kee-AL-is). It's located under the biceps and helps the biceps to flex the elbow.

Your deltoid and pectoral muscles get sensation from nerves in your armpit. Blood is supplied to the front of your upper arm through the *brachial artery* and to the rear of your upper arm by the *profunda brachii artery*.

Here are the major muscles of the upper arm, along with the major arteries that supply blood to the entire arm.

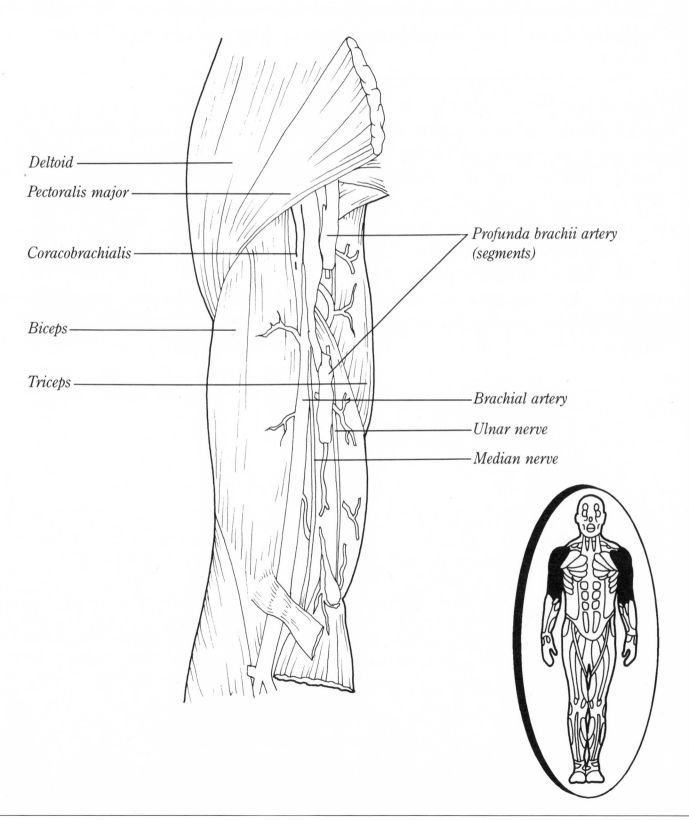

Deltoid

Pectoralis major

Coracobrachialis

Biceps

Triceps

Profunda brachii artery
(segments)

Brachial artery

Ulnar nerve

Median nerve

The Truth about Your Funny Bone

Your armpit area is called the *axilla*, and you can see from this drawing that there's a lot going on under there. The *axillary artery* supplies blood to the muscles of the upper arm and chest, which border the axilla.

At the front of the axilla are *pectoralis major* and *pectoralis minor* muscles. Behind these muscles is the *serratus anterior*. It anchors the shoulder blade to the wall of the chest.

When the axillary artery passes out of the axilla, it is called the *brachial artery* of the arm. Three nerves pass by the axillary artery: the *musculocutaneous* (MUS-kyu-lo-kyu-TAY-nee-us) *nerve*, the *median nerve*, and the *ulnar nerve*.

Your musculocutaneous nerve supplies feeling to your biceps, *coraco-brachialis* (KOR-ah-ko-BRAY-kee-al-is), and *brachialis* muscles.

The median nerve supplies feeling to muscles in your forearms and hands.

Your ulnar nerve supplies sensation to some forearm muscles and most of the small muscles of your hand. When you bang your "funny bone," you're actually irritating the ulnar nerve. And when that happens, it doesn't feel so funny.

The *radial nerve* is behind the axillary artery. It supplies motor control to your triceps and to the muscles at the back of your forearm.

PINS AND NEEDLES

If you sit or sleep in one position for too long, you might make your leg or arm "go to sleep." That numb feeling happens because the blood flowing to the nerves of your leg or arm slows down.

The nerves, which let you move your arm or leg, don't get enough oxygen, so they can't do the job they're supposed to do. When you straighten up, the blood flow returns and the nerves go back to work. That's when you get that pins and needles feeling.

Blood is supplied to each upper arm and each side of the chest by the axillary artery. It's shown here among many important nerves and muscles.

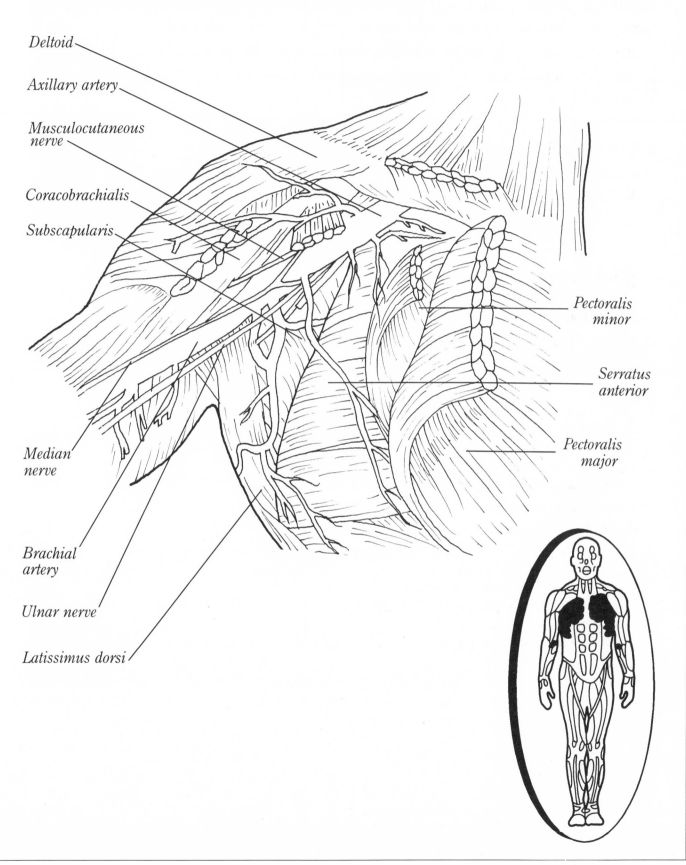

Deltoid

Axillary artery

Musculocutaneous nerve

Coracobrachialis

Subscapularis

Pectoralis minor

Serratus anterior

Pectoralis major

Median nerve

Brachial artery

Ulnar nerve

Latissimus dorsi

Tiny Tubes and Large Veins

Just below the surface of your skin, as well as in your muscles and organs, lies a network of tiny tubes called *lymph vessels*. Here you can see some of the lymph vessels of the chest area.

The three lymph nodes in the *axilla*, or *armpit*, are especially important in women. The breast tissue is located just beneath the skin of the chest. If a woman has breast cancer, some of the cancer cells might break away from her breast and be filtered out by the *axillary lymph nodes*. In this way, a second cancerous mass may start to grow in her armpit.

The surface veins of the arm are especially noticeable. The *cephalic vein* runs up the front of your arm to a groove between the *deltoid* and the *pectoralis major muscle*. At the elbow, it connects with the *basilic vein* through the *median cubital vein*.

If you look at the inside of your elbow, you may be able to see the median cubital vein clearly. This vein is the one most often used when people donate blood to the Red Cross. That's because it's a large vein that lies very close to the surface, and there aren't many pain nerve fibers near it.

This picture shows the lymph system of the chest and arm. The lymph system drains fluids from tissues and protects against infection.

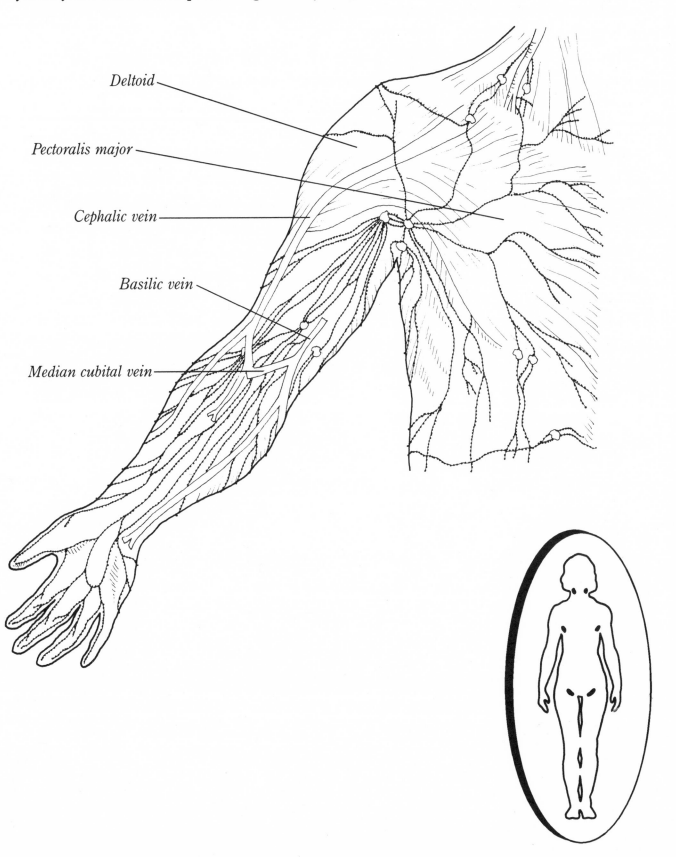

Deltoid

Pectoralis major

Cephalic vein

Basilic vein

Median cubital vein

An Armload of Names

These drawings show the many muscles of your forearm. It may seem complicated, but once you learn how muscles are named, it will make sense to you.

A muscle that bends or flexes a joint is called a *flexor*; a muscle that straightens a joint is called an *extensor*.

The rest of the muscle's name refers to its location, shape, or what part of the body it works on.

For example, *flexor digitorum profundus* means deep (profundus) flexor of the fingers (*digit* means fingers), and *extensor carpi radialis longus* means long extensor of the wrist (*carpi* means wrist) on the side of the radius bone.

The *pronator quadratus* (four-sided pronator) turns the back of your hand to the front (palm down). This motion is called pronation. The *supinator* turns the palm of your hand to the front (palm up). This action is called supination.

The left drawing shows the deeper flexors. All your forearm and hand flexors are at the front of your forearm. The *flexor carpi radialis* flexes the wrist on the radius side. You can see the long tendons of these muscles in the drawing, and you can also see them at your wrist when you make a fist and flex your hand.

The drawing on the right shows the extensors, which are all at the back of your forearm. The *brachioradialis* (BRAY-key-o-RAY-dee-al-is) muscle is the exception. It works with the biceps and brachialis muscles to flex your arm at the elbow.

The *extensor carpi radialis brevis* is the short (*brevis* means short) extensor of the wrist, and the *extensor carpi ulnaris* is the wrist extensor on the ulna side.

The *extensor digiti minimi* means the extensor of the smallest finger, and *extensor pollicis brevis* means short extensor of the thumb (*pollicis* means thumb). Can you figure out what *extensor pollicis longus* means?

The *abductor pollicis longus* abducts or moves the thumb away from the other fingers. The *adductor pollicis* is one of the four muscles in the fleshy part of your palm at the base of the thumb, and it moves your thumb toward your other fingers.

These drawings show the muscles and tendons of the forearm and hand. At left is a front view, showing the deep muscles. At right is the back of the forearm, showing the surface muscles.

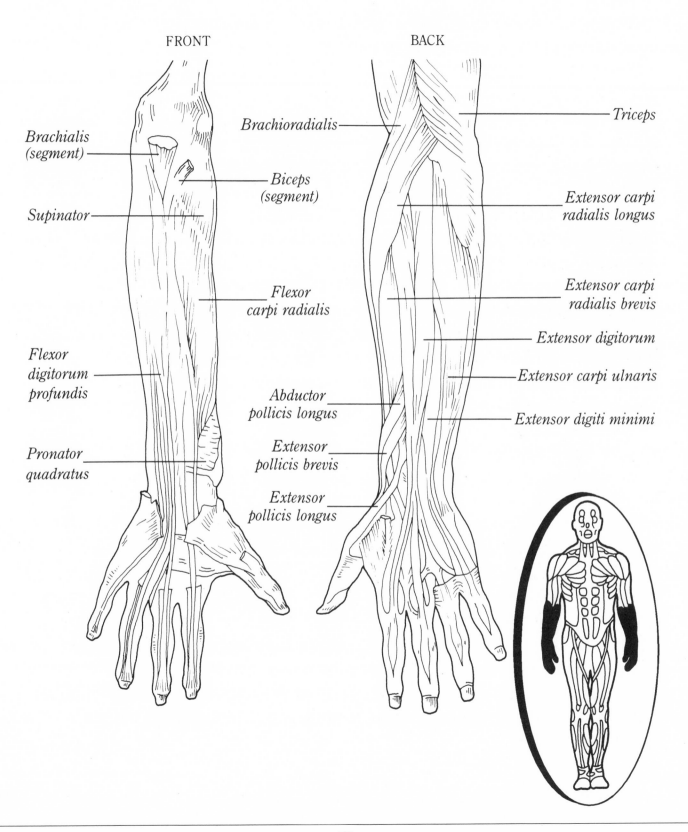

FRONT

BACK

Brachialis (segment)

Brachioradialis

Triceps

Biceps (segment)

Supinator

Extensor carpi radialis longus

Flexor carpi radialis

Extensor carpi radialis brevis

Extensor digitorum

Flexor digitorum profundis

Extensor carpi ulnaris

Abductor pollicis longus

Extensor digiti minimi

Extensor pollicis brevis

Pronator quadratus

Extensor pollicis longus

When you extend your fingers to wave goodbye, you're using the *lumbrical* muscles in your hand.

The palm of your hand has many muscles that control the fine movements of your hand. Your thumb alone has six small muscles so it can bend and wiggle.

The thumb muscles are the *opponens pollicis*, the *abductor pollicis brevis* and *longus*, the *flexor pollicis brevis* and *longus*, and the *adductor pollicis*.

The opponens pollicis lets you fold your thumb in front of your other fingers. The abductor pollicis brevis works with the abductor pollicis longus to pull your thumb away from your other fingers. The flexor pollicis brevis works with the flexor pollicis longus to flex your thumb. The adductor pollicis pulls your thumb toward your other fingers.

Your little finger has three similar muscles, but no muscle to move it across your palm. You can only see two of the muscles in this drawing—the *flexor* and *abductor digiti minimi muscles*. The *opponens digiti minimi* is under the flexor. If you haven't guessed, *digiti minimi* means "little finger."

Your wrist has a flat ligament running across it called the *flexor retinaculum*. Under it run the tendons of the muscles in front of the forearm. This area underneath is called the *carpal tunnel*.

The palm of your hand has many small muscles that let you make gentle finger movements to play an instrument, signal for attention, or tickle someone.

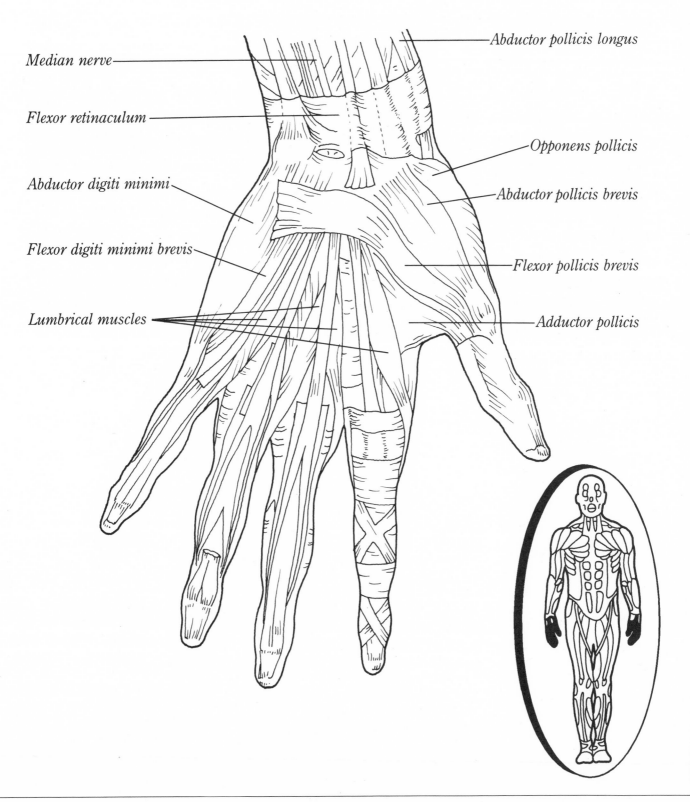

Median nerve

Flexor retinaculum

Abductor digiti minimi

Flexor digiti minimi brevis

Lumbrical muscles

Abductor pollicis longus

Opponens pollicis

Abductor pollicis brevis

Flexor pollicis brevis

Adductor pollicis

Putting Your Best Foot Forward

Your feet are much like your hands, except that your foot bones don't move as freely. But that's O.K. Feet weren't built to be graceful. They are designed for bearing weight, walking, running, and keeping your balance.

Each of your feet may have to absorb pressures of more than a ton per square inch! The bones and ligaments of your feet spread this force and send it efficiently through the tibia and femur in your legs.

At the bottom of your toes are seven *tarsal bones*: the *talus, calcaneus, cuboid, navicular,* and three *cuneiforms*. The talus, or ankle bone, forms the ankle joint with the tibia, medial, and lateral malleolus.

The calcaneus, or heel bone, is the largest and strongest foot bone. A pad of fat between this bone and the skin of your heel helps absorb shock.

The navicular is shaped like a little boat, and the cuboid is cube-shaped. The three cuneiforms are wedge-shaped and lie between the navicular and the first three metatarsal bones.

The sole of your foot is a complex arrangement of bones, muscles, ligaments, nerves, and blood vessels. Your *arches* help your feet absorb the shock of walking and running. They also help you keep your balance.

The bones of your feet may look like the bones of your hands, but your feet are designed to carry your weight.

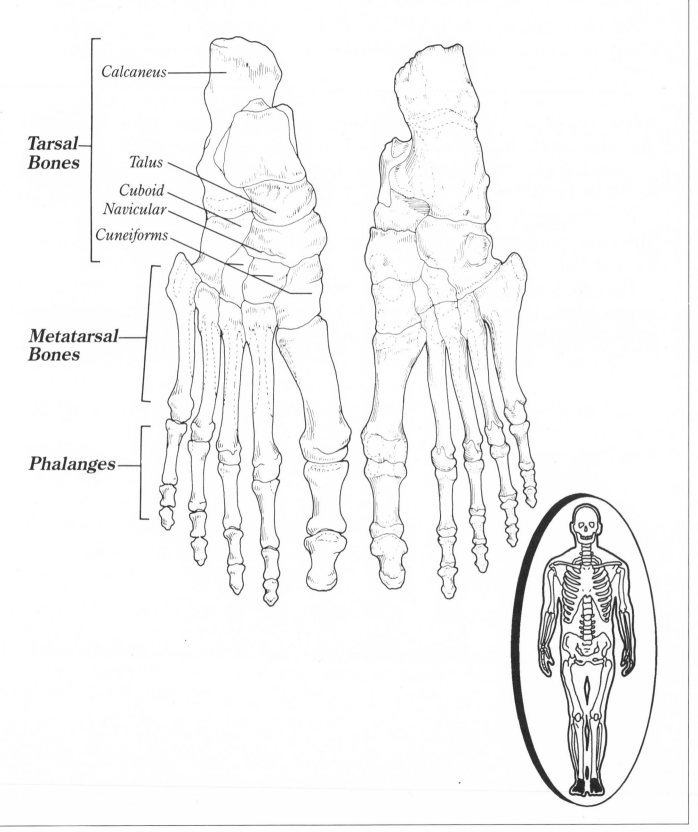

Tarsal Bones

Calcaneus

Talus

Cuboid

Navicular

Cuneiforms

Metatarsal Bones

Phalanges

A Leg Up

Your leg muscles, of course, help you to walk and to keep your balance. They're also among the strongest muscles in your body.

Here are several of the muscles of the back and sides of your leg. The group of muscles on the surface of your calf are the *gastrocnemius* and *soleus* muscles. You can feel them bulge as you flex your foot downward when walking or running.

These two muscles share the same strong tendon, called the *Achilles* (uh-KILL-eez) *tendon*. You can feel your Achilles tendon behind your *calcaneus bone*, or heel bone.

Deep within your calf are four muscles: the *popliteus* (pop-LIT-ee-us), the *tibialis* (tih-bee-AL-is) *posterior*, the *flexor digitorum* (dih-jih-tor-um) *longus*, and the *flexor hallucis* (ha-LOO-sis) *longus*.

The popliteus bends your knee joint. The tibialis posterior flexes your foot downward and inward. You use your flexor hallucis longus to bend your big toe and your flexor digitorum longus to bend the other four toes. Your big toe needs a muscle of its own because you push off with your big toe every time you take a step.

Your knees need many powerful muscles to bend and flex them and also to support your weight. In this drawing, you can see the *biceps femoris*, or *hamstring muscles*; the *semimembranosus*; and the *semitendinosus*. Also shown is the *plantaris*, a very weak muscle that some people don't even have.

On the outside of your lower leg are the *peroneus* (puh-RO-nee-us) *longus* and *peroneus brevis*. Both muscles flex your foot downward, and they also turn your foot outward. You can feel the tendon of peroneus longus above and behind your ankle bone when you flex your foot downward against the floor.

The strongest muscles in your body are in your legs. Part of the gastrocnemius muscle, the strongest muscle in your calf, has been left out of this drawing so you can see powerful muscles beneath it.

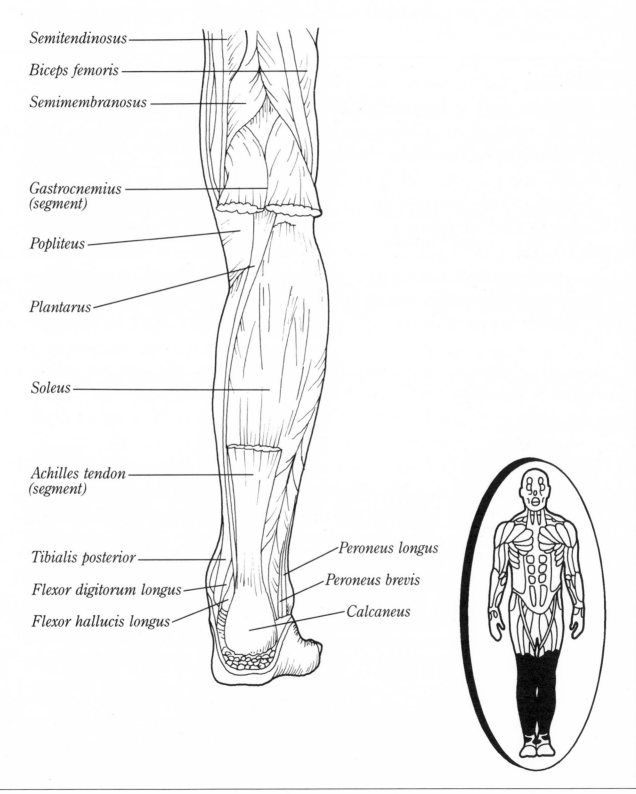

Semitendinosus

Biceps femoris

Semimembranosus

Gastrocnemius
(segment)

Popliteus

Plantarus

Soleus

Achilles tendon
(segment)

Tibialis posterior

Flexor digitorum longus

Flexor hallucis longus

Peroneus longus

Peroneus brevis

Calcaneus

Your leg muscles need a good circulatory system to supply them with blood. The *femoral* (FEM-uh-rul) *artery* and its branches supply all the blood to the leg, and the *femoral vein* and its branches carry the blood from the leg. These blood vessels lie deep within your legs. Even so, you may be able to feel the pulse of the femoral artery where your leg meets your abdomen.

A large surface vein branches from the femoral vein. It's called the *great saphenous* (SAF-uh-nis) *vein*. It has branches all along the surface. They meet with branches of the *small saphenous vein* (also a branch of the femoral vein) at the rear of your lower leg.

You can see a major branch of the femoral artery, called the *popliteal artery*, and its two major branches in the drawing on the right. The popliteal artery supplies blood to the muscles and skin of your rear thighs, your knees, and your legs.

The two branches of this artery are the *anterior tibial artery*, which supplies blood to the front of the lower leg and to the top of your foot; and the *posterior tibial artery*, which along with the *peroneal artery* supplies the rear of your leg and the sole of your foot.

USED PARTS CAN SAVE LIVES

When a doctor takes an organ from one person and puts it into another person, it's called a transplant. Transplanting organs from the recently deceased is one way that doctors help their patients live longer.

Transplants aren't easy, and doctors have to watch for problems. The main problem is that the body might treat the transplanted organ like a germ and try to kill it.

Sometimes a part is transplanted from one area of the body to another. When a person gets a coronary bypass operation, veins are removed from the leg and transplanted to the heart to help restore circulation.

The drawing on the right shows the major veins at the front of the leg. The drawing on the left shows the major arteries at the back.

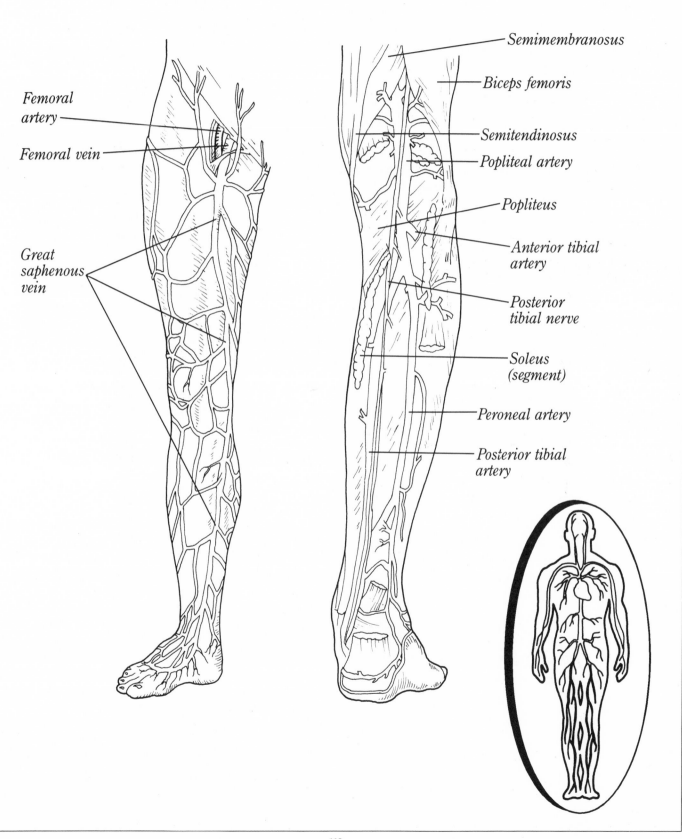

Femoral artery

Femoral vein

Great saphenous vein

Semimembranosus

Biceps femoris

Semitendinosus

Popliteal artery

Popliteus

Anterior tibial artery

Posterior tibial nerve

Soleus (segment)

Peroneal artery

Posterior tibial artery

Blood traveling in the veins of your arms and legs moves against gravity. To keep the blood flowing in the right direction, the veins in your limbs are lined with valves.

The next time your doctor takes your blood pressure, ask him or her to point out some of the valves in the surface veins of your arm. They look like bumps.

The drawing on the right shows the *great saphenous vein*. If a person works standing in one place for long periods of time, the valves in the great saphenous vein may fail. When this happens, gravity causes blood to pool in the vein and its branches. The veins enlarge, causing a painful condition known as *varicose veins*.

A surgeon can remove varicose surface veins without any ill effect because deeper veins can carry the blood just as well.

You can also see the *inguinal lymph nodes* in this drawing. These nodes filter lymph from your leg and help prevent infections from traveling to the rest of the body.

The surface veins of your arm include the *cephalic* and *basilic veins*. They're connected by the *median cubital vein* at the busy intersection at the crook of your elbow. It's a good place to take blood for tests or donations.

Also at this spot, just below the median cubital vein, is the *bicipital aponeurosis* (bi-SIP-ih-tul AP-eh-nyu-RO-sis). This broad, flat ligament connects your biceps to the connective tissue of the flexor muscles of your forearm. It lies on top of the brachial artery. That's the artery that doctors listen to when they take your blood pressure.

On the right is a picture of the great saphenous vein and the lymph vessels of the leg. The drawing on the left shows the major veins of the arm.

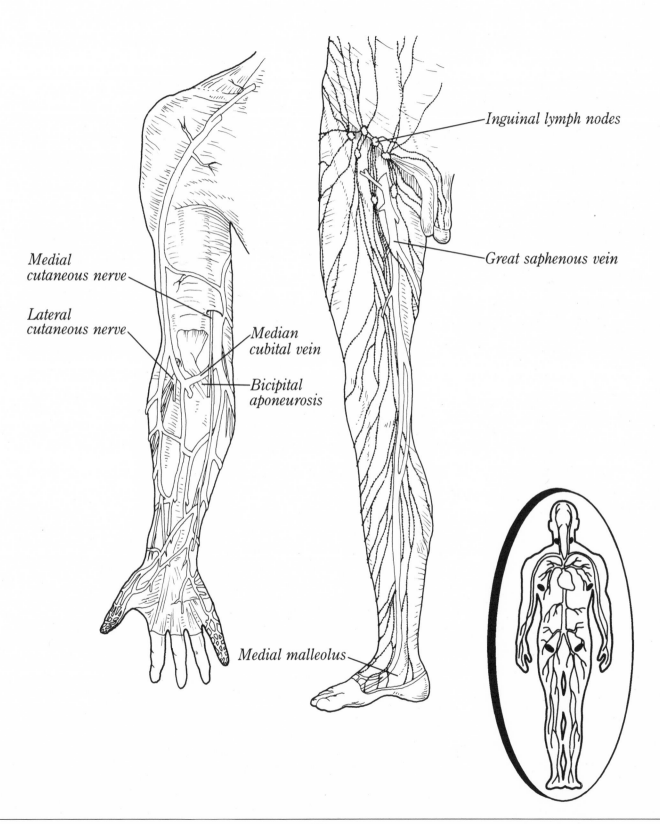

Inguinal lymph nodes

Great saphenous vein

Medial
cutaneous nerve

Lateral
cutaneous nerve

Median
cubital vein

Bicipital
aponeurosis

Medial malleolus

Some Nerve!

The largest nerve in your body is the *sciatic* (sy-AT-ik) *nerve*. With its branches, it supplies motor control to all the muscles at the rear of your thighs, your lower legs, and your feet. It travels under your *piriformis muscle* at your buttocks, down the back of your leg behind your knee. There it branches into the *common peroneal nerve* and the *tibial nerve*.

The common peroneal nerve travels to the outside of your knee and supplies motor control to the muscles on the outside and front of your lower legs. It's not well protected and can be easily injured by a blow to the outside of the knee. The tibial nerve supplies motor control to the muscles of your calves and feet.

Your buttock muscles get their motor control from other nerves. The *gluteus medius* and *gluteus minimus* muscles are supplied by the *superior gluteal nerve*. The *gluteus maximus* is supplied by the *inferior gluteal nerve*.

Here are the major nerves and muscles of the back of the right leg.

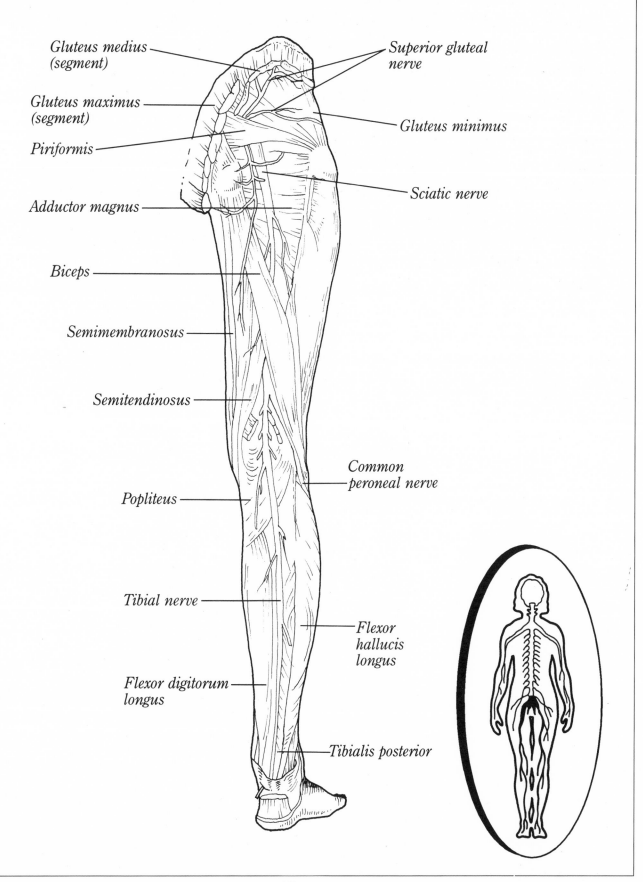

Gluteus medius (segment)

Superior gluteal nerve

Gluteus maximus (segment)

Gluteus minimus

Piriformis

Sciatic nerve

Adductor magnus

Biceps

Semimembranosus

Semitendinosus

Common peroneal nerve

Popliteus

Tibial nerve

Flexor hallucis longus

Flexor digitorum longus

Tibialis posterior

Nerves have two jobs. They carry messages from your brain that tell your body how to move, and they tell your brain if you're hot, cold, or being tickled.

The nerves near the surface of your skin are called *cutaneous* (kyu-TAY-nee-us) *nerves*. They supply the sensations of heat, cold, and touch. They also control the surface blood vessels. When you're cold, the cutaneous nerves close the blood vessels, keeping in your body heat. When you're hot, the cutaneous nerves open the surface blood vessels, letting some body heat escape.

The skin of your lower leg and the top of your foot are supplied by the ends of two branches of the *peroneal nerve*: the *superficial peroneal nerve* and the *deep peroneal nerve*.

The main nerves in your feet are the *medial plantar nerve* and the *lateral plantar nerve*. They are both branches of the tibial nerve.

This is a view of the right leg and the bottom of the foot, showing nerves that supply the skin and some muscles.

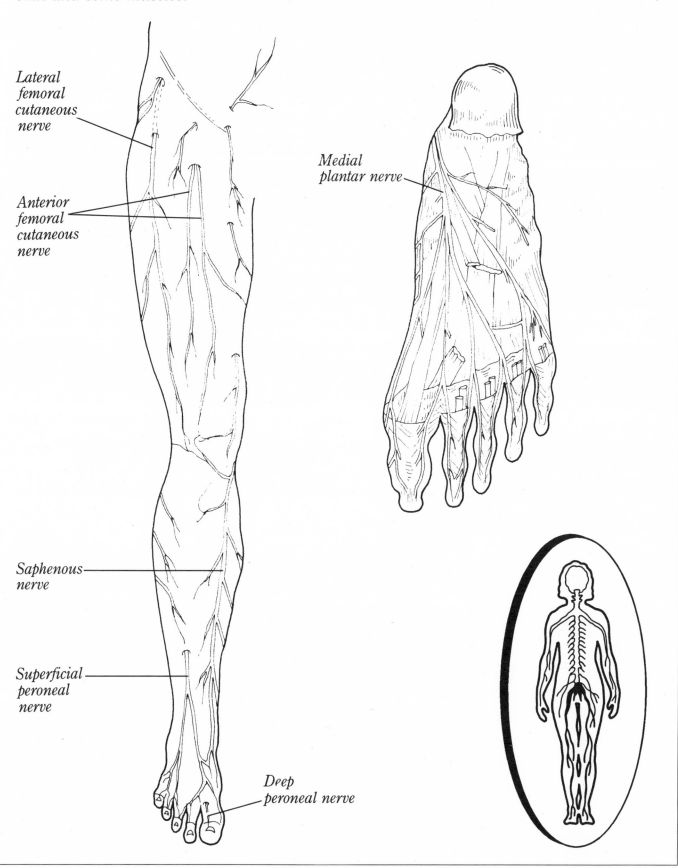

Lateral
femoral
cutaneous
nerve

Anterior
femoral
cutaneous
nerve

Medial
plantar nerve

Saphenous
nerve

Superficial
peroneal
nerve

Deep
peroneal nerve

Your *skin* is the largest organ of your body! To protect you from overheating, it sweats and blushes. To keep you from getting too cold, your skin can close down its blood supply, keeping your body heat inside.

Skin has two layers. The outer layer is called the *epidermis* (eh-pih-DER-mis). It's a barrier that keeps disease out and water in. The inner layer of skin, the *dermis*, contains your skin's blood and nerve supply.

The epidermis is constantly renewed. You shed flakes of skin every day as you wear away this outer layer. In areas of your body where there's friction, such as on the soles of your feet, the skin forms a thick layer of epidermis called a callous.

Your hair, nails, and sweat glands are special structures of your skin.

Every hair has two parts: the bulb and the shaft. The bulb is in the dermis, and the shaft extends outward to the surface. Tiny muscles, called *erector pili muscles*, are attached to the hair shaft. They make your hair stand on end when you're frightened or cold. Oil glands coat the hair shaft with oil.

Nails may help make your sense of touch more sensitive by providing a rigid backing. Nails grow faster in the summer, and your fingernails grow about four times faster than your toenails. The normal color of nails is pink. Blue nails mean that not enough oxygen is reaching your fingers. This can be caused by disease, poisons, or just by being in cold water too long.

Sweat glands are located all over your body. They secrete mostly water to help keep you cool in the heat.

A closeup, side view of skin, showing a hair follicle. Your outer layer of skin is constantly flaking off as a new layer grows back.

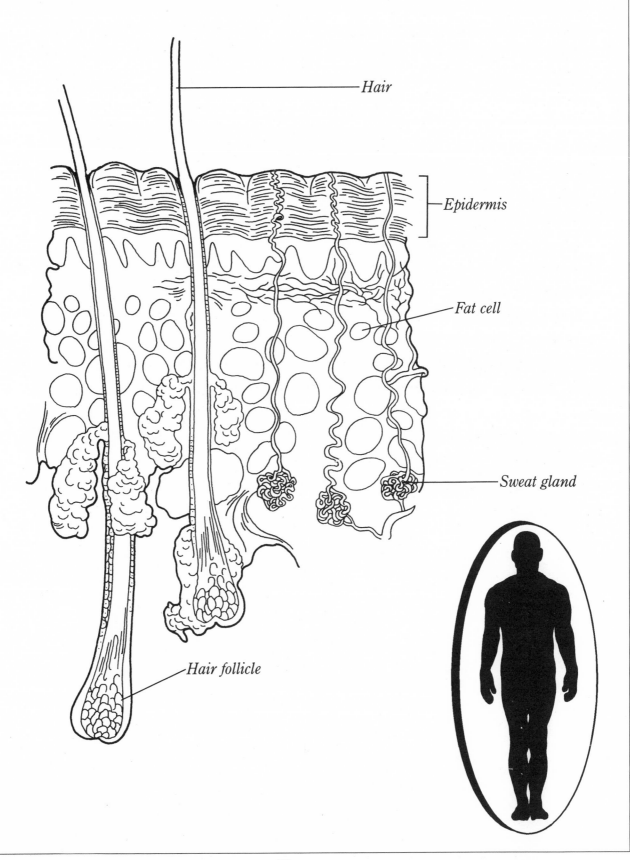

Hair

Epidermis

Fat cell

Sweat gland

Hair follicle

The reproductive system is a group of organs that creates new life. Males and females have different reproductive systems that work together to make babies.

This drawing is a side view of the body showing parts of the male reproductive system.

The *testes* (TES-teez), or *testicles*, are where *sperm cells* are produced. Sperm cells are messengers. They carry half the information needed to make a baby. The rest of the information is in the egg cell, which is part of the woman's reproductive system.

Sperm cells have long tails that propel them through the female reproductive system to finally meet and fertilize the egg cell.

The testes also produce male hormones. These hormones are called *androgens* (AN-dro-jinz). They're responsible for the changes in a boy's body as he becomes a man. For example, androgens determine when a young man's beard begins to grow and when his voice will get deeper.

Sperm travel through a tube in the spermatic cord. At the point where the two tubes join (one from each testicle), the walnut-sized *prostate gland* and the *seminal vesicles* add fluid to the sperm.

This fluid, along with the sperm cells, is ejected through the *urethra*, a tube in the *penis*. Although the fluid contains millions of sperm, only one single sperm cell is needed to fertilize an egg.

In the male, the urethra has another use besides reproduction. It's also part of the urinary system. *Urine* is your body's liquid waste. It is formed in the *kidneys* and travels down long tubes, called *ureters*, to the *bladder.* Urine is temporarily stored in the bladder until it is eliminated through the urethra.

This is a side view of a man's reproductive system. Also shown are the bladder, rectum, and the tip of the spine.

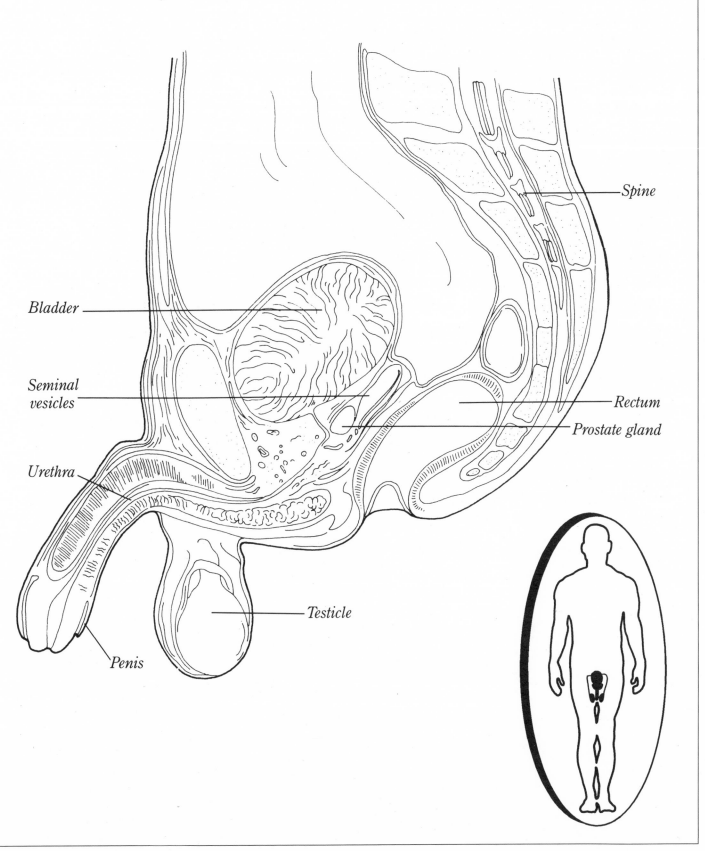

Spine

Bladder

Seminal
vesicles

Rectum

Prostate gland

Urethra

Testicle

Penis

The female's reproductive system is much different from the male's. A woman's reproductive organs are designed to receive sperm from the man; provide a place for the sperm and the *ovum*, or *egg cell*, to meet; and nourish and protect the developing baby.

The ovum is produced in a pair of glands called *ovaries*. An ovary usually releases a single ovum each month. If two or more ova (plural for ovum) are released at one time, twins or triplets may result.

The sperm cells provided by a man swim through an opening called the *cervix* (SER-vix), into the *uterus* (YU-tur-iss), and then into the *fallopian* (fal-O-pee-in) *tubes*.

When released from the ovary, the ovum passes through the fallopian tube, where it may meet the sperm cells. Only one sperm can fertilize an ovum at one time. When a sperm cell and the ovum combine, fertilization takes place, and the woman becomes pregnant.

When fertilization occurs, the cell that results from the joining of the ovum and the sperm cell is called the *zygote*. The zygote divides, creating a bundle of cells. These cells travel into the *uterus*, where they continue to grow.

Besides producing ova, the ovaries produce female hormones that control the monthly release of the ovum. These hormones also prepare the uterus for the possibility of pregnancy by increasing the thickness and blood supply to the uterus's inner lining.

This side view of a woman's pelvic region shows the position of the uterus along with the bladder, rectum, and the tip of the spine. The uterus can expand to many times its size to hold a developing baby.

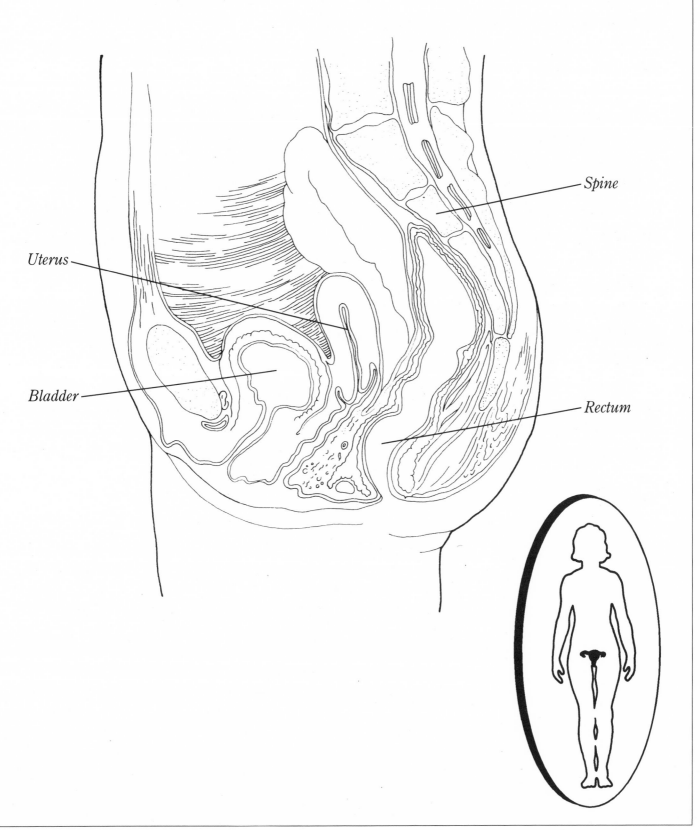

Spine

Uterus

Bladder

Rectum

After fertilization, the single-celled *zygote* rests about one day preparing to divide. The zygote divides into two cells, then the two cells become four, the four become eight, the eight become 16, and so on until the cells form a hollow ball. We call that ball of cells the *embryo*.

By the end of the first week, the embryo attaches to the inner lining of the uterus. The arteries of the lining deliver food and oxygen to the developing embryo while the veins remove the embryo's wastes.

During the second week, three different cell layers form. These layers give rise to all the organs of the body.

The outer layer becomes the skin, nails, hair, sensory organs, and the nervous system. The inner layer develops into the liver, lungs, and the digestive system. The middle layer will become bone, muscle, kidneys, and the circulatory system.

The nervous system begins to form at 2½ weeks. The embryo's heart starts beating at about 3½ weeks, pumping blood for the rest of its life.

The *placenta* is the region where food, oxygen, and wastes are exchanged between the embryo's and the mother's blood. The *umbilical cord* attaches the embryo to the placenta. It has two arteries that carry blood to the placenta, and a vein that carries blood back to the embryo.

The embryo floats in a fluid-filled cavity called the *amniotic* (am-nee-OT-ik) *sac*. This sac helps cushion the embryo from jolts and shocks.

By the end of three months, the embryo is about three inches long and has developed almost every basic structure. It is now called a *fetus*.

In the remaining six months before it is born, the fetus grows larger and its organs more complex.

At birth, it will be a complete human being with perfectly working organs. All that from just a single cell!

This drawing shows the circulation of blood between a mother and her unborn child. When the baby is born, the umbilical cord is cut. The scar that's left becomes the belly button.

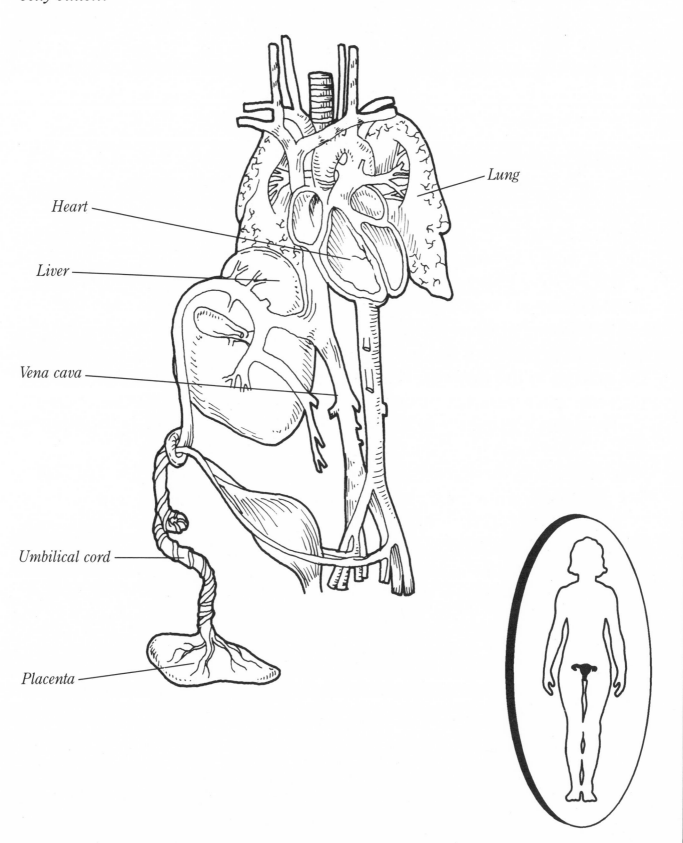

Heart

Lung

Liver

Vena cava

Umbilical cord

Placenta

The Running Press Start Exploring™ Series
Color Your World

With crayons, markers and imagination, you can re-create works of art and discover the worlds of science, nature, and literature. Each book is $9.95 and is available from your local bookstore. If your bookstore does not have the volume you want, ask your bookseller to order it for you (or send a check/money order for the cost of each book plus $2.50 postage and handling to Running Press).

THE AGE OF DINOSAURS

by Donald F. Glut

Discover new theories about dinosaurs and learn how paleontologists work in this fascinating expedition to a time when reptiles ruled the land.

ANCIENT EGYPT

by Peter Der Manuelian

Text and detailed illustrations teach about the geography, culture, daily life, architecture, and kings and queens of ancient Egypt.

ARCHITECTURE

by Peter Dobrin

Tour 60 world-famous buildings around the globe and learn their stories.

BALLET

by Trudy Garfunkel

An ideal introduction to the world of ballet, with text and illustrations covering ballet's history, costumes, and dancers.

BULFINCH'S MYTHOLOGY

Retold by Steven Zorn

An excellent introduction to classical literature, with 16 tales of adventure.

THE CIVIL WAR

Blake A. Magner

Text and 60 ready-to-color illustrations detail the Civil War's sequence of events and show how the great leaders of the North and South shaped the nation.

FORESTS

by Elizabeth Corning Dudley, Ph.D.

Winner, *Parents' Choice* "Learning and Doing Award"

The first ecological coloring book written by a respected botanist.

GRAY'S ANATOMY

by Fred Stark, Ph.D.

Winner, *Parents' Choice* "Learning and Doing Award"

A voyage of discovery through the human body, based on the classic work.

MASTERPIECES

by Mary Martin and Steven Zorn

Line drawings and lively descriptions of 60 world-famous paintings and their artists.

OCEANS

by Diane M. Tyler and James C. Tyler, Ph.D.

Winner, *Parents' Choice* "Learning and Doing Award"

An exploration of the life-giving seas, in expert text and 60 pictures.